New Roots For Agriculture

Wes Jackson

Friends of the Earth
San Francisco

Published in cooperation with
The Land Institute

The Land Institute
Route 3
Salina, Kansas 67401

Friends of the Earth
Principal Offices

124 Spear Street
San Francisco, California 94105

530 Seventh Street, S.E.
Washington, D.C. 20003

72 Jane Street
New York, New York 10014

Library of Congress Catalog Card Number 79-56913

ISBN: 0-913890-38-3

Cover design by Bill Oetinger
Cover illustration by Bobbie Lively
Designed by Nancy Austin
Text illustrations by Marty Bender

Trade sales and distribution by Friends of the Earth,
124 Spear Street, San Francisco, CA 94105.

Lovingly dedicated to my mothers:

Mother earth, who has been here some four billion years now;

Our grassland mother who coevolved with our species;

Ceres, the goddess of agriculture, as misdirected and outright destructive as she has been;

My biological mother, Nettie Stover Jackson, who is still living on the farm on which she was born in 1894;

And Dana, the mother of our children, who has been my partner and best friend for a quarter century.

—*W. J.*

Contents

Preface

Most analyses of problems *in* agriculture do not deal with the problem *of* agriculture. Most talk is about problems with the way farming is done (problems *in* agriculture), not of the threat *of* agriculture to the biosphere itself. Some people warn of the domination of agribusiness, scorn corporate farming and stress the need for land reform. In response, we are now beginning to see support for the small family farm, but so far the small organizations that promote it have almost no political power. This is a problem *in* agriculture.

Farmers who have managed to avoid being squeezed off the land perceive the agricultural problem as one of necessary, but large and scary, capital investment coupled with a helplessness in regulating prices. Environmentalists worry about chemical farming, an extension of their traditional anxiety about industrial pollution. Neither are problems *of* agriculture, nor is everyone's worry about the loss of prime land to urbanization.

All the problems *in* agriculture are legitimate. They require thought, passion and action a hundred times greater than they now receive.

Now and then we see soil loss as a major threat. But even here, we usually fall short of bringing the entire problem into sharp focus. Rather the associated thoughts are usually directed toward some naughty farmers (seen in the abstract) who are not maintaining their terraces or rotating their crops properly or, worse yet, must be plowing up and down the hill instead of on the contour. In the flat Great Plains wheat country, their counterparts are failing to leave a stubble mulch to minimize blowing dust.

Most people believe there is a right way to *do* agriculture and that failure to *do* it correctly is simply a failure in character. The very *nature of farming* itself is seldom called into question, and the one who does question agriculture itself may be accused of wanting to return to a bow and arrow way of life.

Agriculture is seen potentially as an altogether wholesome enterprise. A color photo of a well-scrubbed, 4-H kid, clasping the rope on an equally well-scrubbed Holstein heifer at the country fair is *always* fit for a September calendar put out by the local feed store. One glance tells that he or she will do better on the farm. Don't the bright eyes radiate "new knowledge" and the rosy cheeks, "wholesome values?"

It isn't just through the kid at the county fair that the notion of the inherent wholesomeness of agriculture gets a boost. The appropriateness of till agriculture is firmly implanted in all civilized peoples. At the United Nations there is a huge statue of a man full of purpose and muscle bent to the task of beating a sword, which does evil of course, into a plowshare, which everyone knows will do good. The developer of a new idea may be described as having "plowed new ground." Saul, the first king of Israel, was annointed in the field where he had been plowing with oxen, suggesting at least a left-handed endorsement of till agriculture from the Almighty himself. The concept of till agriculture is interwoven in our metaphors and symbols. Yet the plowshare may well have destroyed more options for future generations than the sword.

This book calls essentially all till agriculture, almost from the beginning, into question, not because sustainable till agriculture can't be practiced, but because it isn't and hasn't been, except in small pockets scattered over the globe. So destructive has the agricultural revolution been that, geologically speaking, it surely stands as the most significant and explosive event to appear on the face of the earth, changing the earth even faster than did the origin of life. Volcanoes erupt in small areas, and mountain ranges require so long in their uplift that adjustments to changing conditions by the life forms are smooth and easy. But agriculture has come on the global scene so rapidly that the life-support system has not had time to adjust to the changing circumstances. In this sense, then, till agriculture is a global disease, which in a few places has been well-managed, but overall has steadily eroded the land. In some areas, such as the U.S., it is advancing at an alarming rate. Unless this disease is checked, the human race will wilt like any other crop.

Agriculture has been given every chance to prove itself as *a viable experiment for continuously sustaining a large standing crop of humans*. Its failure to do so is difficult to comprehend because since Jamestown, each decade, if not each year, we North Americans have harvested more and more food. In spite of all our scientific and technological cleverness of recent decades, *not one significant breakthrough* has been advanced for a truly *sustainable* agriculture that is at once *healthful* and sufficiently *compelling* to be employed by a stable population, let alone an exploding one. Even when we do think deeply about the problem, we are inclined to accept the eventual decline of agriculture as being in the nature of a tragedy in drama—inevitable.

In *The Unsettling of America,* Wendell Berry raises our sights on the agricultural problem by dealing with most of the problems *in* agriculture. Berry sees that the problem basically comes from a failure of the human spirit. I discuss both the religious dimension and the importance of regarding the farm as a hearth rather than a food factory but after an examination of the many areas in which we have failed over the centuries to adequately deal with the agricultural problem, I offer an additional consideration. I think agriculture needs a technical fix—a bio-technical fix. I am not entirely comfortable with such a proposal—not because I don't believe in its promise, but because of what technical fixes promote even if they are of a "bio" variety. They feed the modern day zealot who sees all of our problems to be materially or technically solvable. Such an attitude prevents us from facing up to the deeper moral and spiritual problems Berry so eloquently considers.

Nevertheless, because of advances in biology over the last half-century, I think we have the opportunity to develop a sustainable agriculture. This bio-technical fix would be based on mixed perennial seed-producing plants that would make it easier for humans to solve many problems *in* agriculture at once.

As civilizations have flourished, many upland landscapes which supported them have died, and desert and mudflat wastelands have developed. However, the civilizations passed on accumulated knowledge, and we can say without exaggeration that these wastelands are the price paid for the accumulated knowledge. In our century this knowledge has grown enormously, and on the balance, it seems more arrogant and sinful to ignore this knowledge than to recognize its restorative potential. I think we have the chance to develop a truly sustainable food supply, something most of the globe has not enjoyed since we stepped onto the agricultural treadmill.

There are numerous success stories in agriculture, both of individuals and cultures, that have managed to maintain healthful, productive and sustainable farms. Most of the northern European cultures and Japan have farms that are maintained in a seemingly sustainable way. But as we look at the success stories, we discover that a complex of factors exists, including the nature of the rainfall, the nature of the cropping system, the nature of the soils, and the nature of the culture, which combine in unique ways to promote a positively compelling sustainable agriculture. Even so, neither northern Europe nor Japan comes close to feeding itself. And the number of individuals or cultures that practice a sustainable agriculture that is positively compelling—that is for reasons other than fear of starvation or group pressure, as in China—is small indeed.

To suggest that the solution to the agricultural problem simply requires following the example of the ecologically correct around us today is a little like suggesting that if more people were like citizen Doe who displays good conduct, no police or military would be needed. Well, both the police and military do exist and both are signs of failure within and of civilization. And so it has always been. But should we not be constantly looking for ways to make them unnecessary? Should we not strive to create an agriculture which makes unnecessary the example of exemplary people within the current agricultural tradition?

I think that if we solve the problem *of* agriculture, we can solve most of the problems *in* agriculture. I think that mixed perennial seed crops can be developed over the next 50 to 100 years, and that they could be sufficiently compelling to be widely adopted.

Wes Jackson
Salina, Kansas
April 12, 1980

Prologue:

Stewards of the Land

One recent June Sunday two friends and I were driving home along a blacktop road through south-central Kansas, Mennonite country. The previous night, and continuing into the morning, much of the state had experienced hard rains, in some places five inches and more. Such storms are not infrequent in Kansas. From earliest childhood, native Kansans, indeed all Great Plains people, are keenly aware of the hair-trigger which stands between drought and instant drenchings that are often accompanied by spectacular lightning displays and high winds.

As we drove through this relatively flat and prosperous Mennonite country, with its tidy fence lines and well-kept houses and farm buildings, we saw roadside ditches and newly opened furrows. They probably contained milo sorghum seeds, but were so full we could not tell whether the crop had germinated or not. The ditches and furrows were not full of water, but of rich black mud—that blackness characteristic of rich prairie soils. This particular landscape has little topographic relief, but what little there is had been accentuated in the past few hours by diagonal washes, five feet wide and more, leading down to the ditches which were now level with the adjacent fields. The little streams of the area were running full and muddy.

A hundred years ago the German-speaking, Russian-born ancestors of these Mennonites had introduced hard winter wheat to the United States and with it the easily-copied cultural practices which eventually gave the Great Plains region its well-deserved reputation as a breadbasket. These farmers, like their close religious relatives, the Amish, believe the highest calling of God is to farm and be good stewards of the soil. Within an agricultural context, they are usually regarded as the most ecologically-correct farmers of any in America. The strong ethic of land stewardship is, without a doubt, largely responsible.

Less than an hour's drive to the east, my friends and I had spent a memorable, leisurely afternoon surrounded by several thousand acres of tall grass prairie country in our state's lovely Flint Hills. We had met other friends and together had botanized, birded and picnicked under a still cloudy but unthreatening sky. Upland plovers were everywhere joining their sounds with the nighthawks, scissor-tailed flycatchers, and meadowlarks. The storm seemed to have immediately invigorated such attractive plants as Showy Evening Primrose, Pale Echinacea, Plains Larkspur, Butterfly Milkweed, and Lace Grass. It was clear on this rich prairie that the rain was being retained long enough in the spongy mass to give the soil a chance to slowly soak it in and then become a reservoir of water for future needs.

In the Mennonite's field, the water had run off, except where it stood idle in puddles. Soil that had become mud was deathly quiet. Even the most casual observer of nature would not fail to see the contrast. The hills are living, and, so long as they are clothed, eternal; the relatively flat lowlands, put to the plow by scarcely three generations of land stewards, are ephemeral.

As we stopped to photograph one severely eroded field, my mind turned to the owner of that field and tried to imagine what was going on in his head on this wet Sunday. First of all, he will have to replant. It will be substantial, he will think, but necessary and affordable. But in this late afternoon, before chore time and evening church services, is he wondering how many more rains like that his fields can take, and is he asking what, after all, is the meaning of land stewardship, which is central to his faith?

A Mennonite at a rodeo is unlikely. Rodeos are wild places; the boisterous sons of ranchers are not known for their piety. But these cattlemen are stewards of the grasslands—though they probably don't think of themselves in that language—and they need only one ethic. It is simple, straightforward and easily taught to their children: "No more than one cow-calf unit to about seven to ten acres, and start moving them off when it's dry." The rancher knows it can rain and blow, and his sons can attend the rodeo, chew tobacco, drink beer, miss church and never mention stewardship, let alone think about its implications. Many ranchers do overgraze and their soil does erode, but even with overgrazing, poor ranching and no ethic, the land fares generally better when grazed than when put to the plow.

For some soil types, under some climatic conditions, a strong stewardship ethic works—but this is the exception rather than the rule.

In the earliest writings we find that the prophet and scholar alike have lamented the loss of soils and have warned people of the consequences of their wasteful ways. It seems that we have forever talked about land stewardship and the need for a land ethic, and all the while soil destruction continues, in many places at an accelerated pace. Is it possible that we simply lack enough stretch in our ethical potential to evolve a set of values capable of promoting a sustainable agriculture?

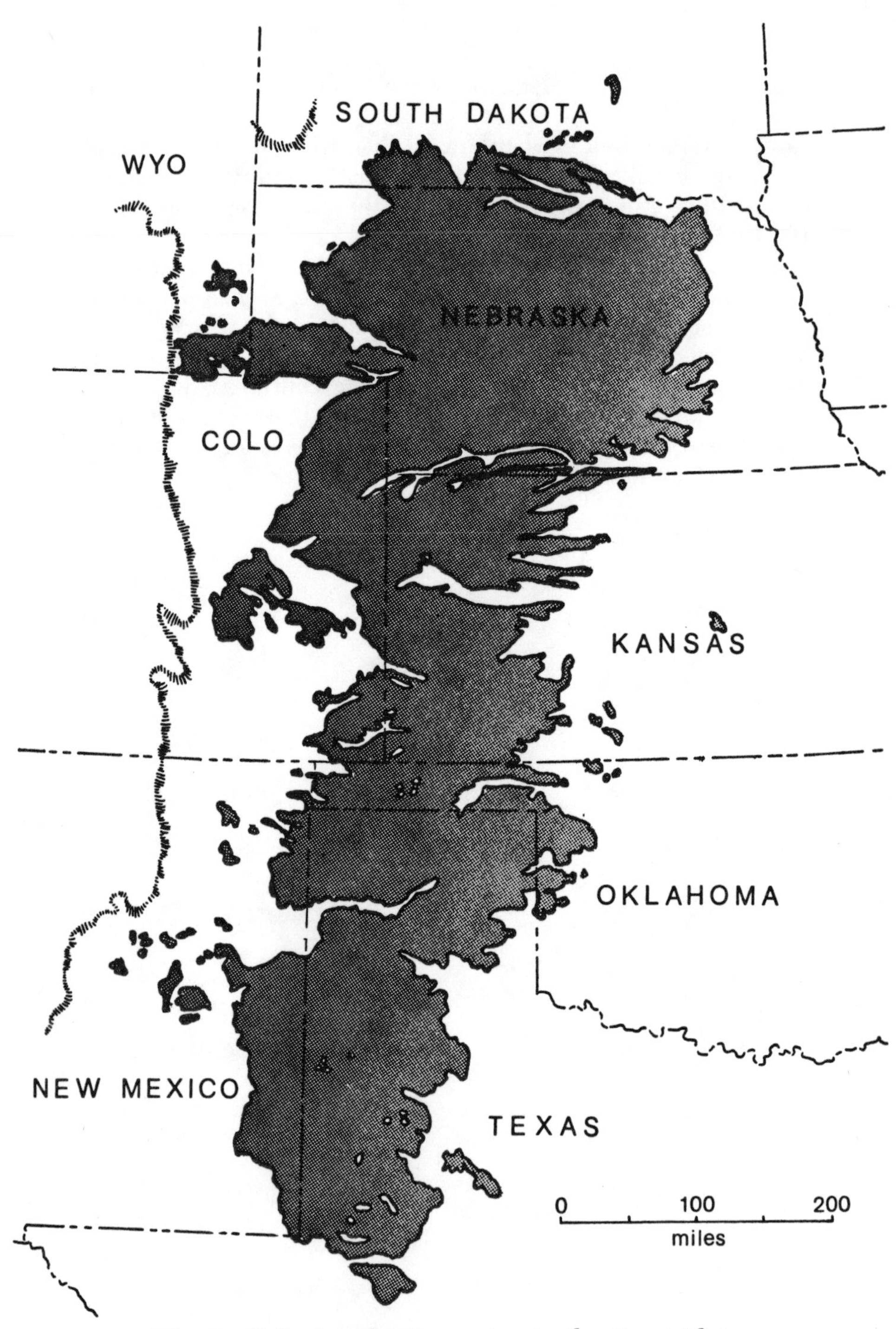

The Ogallala Aquifer Formation in the Great Plains

1
The Earth in Review: The Rise, Role, and Fall of Soil

A quick overview of our planet shows a history that began to get really interesting about 750,000,000 years ago. That is one-sixth of the total age of the earth. The earth spent five-sixths of its time getting set for the explosive emergence of higher life. Some twenty-five of the major phyla around us today appeared then.

For convenience, let us telescope the recent one-sixth into a year, for a quick look at the significant events of this part of the earth's history. We start on January 1. By the fifteenth of March we can see several marine invertebrates and we think we can even see lichens on land. Some time after mid-June there are scorpions crawling about and these newcomers are joined by the first bog plants later in the month.

The lung fishes appear in early July. By late August early reptiles inhabit a landscape dominated by swamp forests, and as we approach September we can see the cone-bearing plants becoming forest trees. In late September, the Auraucarian forests (Norfolk Island Pine and Monkey Puzzle tree are modern descendents of this group) are quickly followed by other seed plants resembling pines.

Sometime in late October we get our first glimpse of flowering plants. A month later it has become obvious that the dinosaurs are headed for extinction. By December 11 some insignificant little mammals with a larger brain-to-mass ratio than the reptiles have become conspicuous, and by a short week later they are the dominant animal group. The mammals have made it. We are all fascinated as we watch the Miocene uplift that creates a rain shadow east of the Rockies, which in turn gives rise to the great North American grasslands. A few days before Christmas we see extensive grasslands in various parts of the planet.

Creatures best described as ape-men appear right after Christmas, and with about thirty hours left in the year, we see a creature which is decidedly human-like, even though it shows little promise at first.

As we watch these creatures closely, various forms develop, most with no future at all; but with less than three hours of the year's last day left (or about 200,000 real years), a creature with a brain almost as large as our own is eking out a livelihood in ecosystems not much different from what we find in many parts of the few wild places left today.

An important system was developing literally under the feet of these diverse life forms. The early dust of the earth was mostly cemented together. It gradually became pulverized by the action of wind and water, plant roots and gravity. The bodies of dead plants and animals were added to this powder. A peculiar type of evolution was under way. This entity teemed with small organisms which secreted chemicals into the powder. Small life forms ingested and egested it, buffered it and burrowed in it. It grew in thickness and began to cover a large area with what we might call "ecological capital." The capital of soil creates "interest" in the form of more soil. This interest then becomes reinvested. Water and wind still carried tons of this capital to the sea to become sedimentary layers, as it always had, but the life forms seemed almost purposefully devoted to retarding this work of gravity. From one point of view, David Brower has humorously suggested, plants and animals were evolved by this soil system to save itself and further its own spread.

A book written in 1905 by Harvard professor Nathaniel Southgate Shaler entitled *Man and Earth* described the soil and water system as an enveloping membrane or film, a *placenta,* through which the Earth mother sustains life. All life, including humans, Shaler suggested, draws life from the sun, clouds, air and earth through this living film. If the placenta is not kept healthy or intact, life above suffers. If healthy, it is a rich, throbbing support system. His message was clear enough: protect the placenta and you protect all Nature's children.

Placenta may not be the best word, for once a birth is complete the placenta is disposed of. And yet Mother Earth is always pregnant with new life and therefore an intact placenta is necessary. Perhaps a better word is matrix. To the biologist a matrix is something within which something else originates or develops. In archaic Latin, *matrix* means a uterus or womb; the word is derived from *mater* or mother. But this is the age of the computer and now the word is associated with the computer. Call it what you will, soil is important not just for land life but for life in the ocean around the continental shelves. In fact, the open ocean is a desert. It would seem as if all life forms—except plants—take this system for granted, regarding it much as they would regard gravity. When humans arrived, they, like the other animals, paid it no special respect.

In the early morning hours of December 31, changes took place on the surface of the earth. Later in the day the human population would explode. But before that, the first glaciers came and the placenta was gouged without mercy. The rubble in their wake was altogether unbecoming, but the placenta persevered; in fact, its speed of growth was increased. One glacier would do to a rock in a year what nosing roots would have done to it in a thousand. The adversity of chilling, grinding glaciers had created a richer life-support system.

Before the glaciers, temperate zone vegetation was draped across the warm northern hemisphere. When the ice age was over, East Asian plants were isolated from their relatives in eastern North America. In Europe, many species retreated to the Mediterranean. Others were isolated by glaciers which descended the east-west European mountain ranges from the north. Finally, East Asia was the richest area in the number of plant species, for the land mass there continued all the way to the equator. Eastern North America was second and Europe, relative to the other two continents, was impoverished. The evolution and extinction of such impressive beasts as the hairy mammoth and the giant bison were like so many bubbles rising and bursting. It was the placenta that was everlasting. We need not worry about the future variety of life on the land. The placenta had stood the test of the glaciers, and had emerged stronger than ever.

And as for *Homo sapiens*? The human species, ten minutes before year's end, was on all major land masses except Antarctica. It was in the next five minutes—from 15,000 to 8,000 years ago—that something critical happened. Gradually, an invisible claw began tearing at the placenta. It wasn't dramatically ruptured as it had been by the ice; there was just a little scratch which failed to heal in the Middle East, and shortly another like it appeared in middle America. The larger the gash, the larger the concentration of people and their handiwork around it. The placenta itself was being ripped away to build civilization. Within three of those last five minutes, the face of the earth was changed. In some places scarcely anything would grow. Scabs—sterile areas or deserts—increased in size owing to human-directed activity. In the last fifteen seconds of the year, the continent of North America was discovered by the Europeans. The great wildernesses of North America disappeared, and the placenta wasted away faster than it had in any other area of the world.

Nearly half of it disappeared in the year's last eight seconds. In the final three seconds, a new stream of oil began to flow throughout the country, and out of it, fossil fuel that had been forming for eight months of our telescoped year, was discovered and was about to be used up in six seconds.

It was now being used not only for transportation, but also as feedstock for chemical fertilizer, in pest control, and in energy for traction in the fields. Clearly a very new thing was happening on earth. Production of living plants was shifting from total dependence on soil to an increasing dependence on fossil fuel. The new reality was clear—agriculture in America was shrinking the placenta, but the decline was obscured by heavy doses of petroleum-based chemical agriculture.

If we were far enough out in space for the planet to seem but the size of an egg, then all the soil, that thin, unique miracle, alive and sustaining life, would, if gathered together in one spot, be barely visible to the naked eye. Built by nature during our telescoped year, half of it lost by man, the self-proclaimed wise one, in a few seconds.

The intensity of the entire agricultural operation can thus be seen as a frantic last attempt to keep alive a rapidly wasting cancer patient. Unless the health of the placenta is restored, a last convulsion will follow, throughout the countryside and around the world.

Review—In Real Time

Seven hundred fifty million years is too much for an individual to grasp, considering the average lifespan of seventy-five years. Seventy-five hundred years is almost too much. But we can speculate that following the ice, humans were sharper and loved their offspring more, and had a somewhat stronger sex drive. The first agricultural revolution began, and by the time most people had moved from hunting to tilling, within about 2,000 years, essentially all our crops and livestock of today had been developed. Civilization has been both the product and promoter of this revolution. If we review these few millenia the way a football coach looks for the turning point in the game his team has lost, we first discover the lesions of soil loss near the Zagros Mountains of western Iran between old Persia and Mesopotamia. A streak spreads westward and to the east and south to include much of India and East Asia. Some humans had learned to accelerate the processes of nature. Draft animals, followed by tractors and hand tools, followed by power-driven machinery, all served as instruments of destruction. The yield of food appears to increase with each improvement, but this only results in more people the next generation. There were some major die-offs, but these die-offs were mere downward anomalies in a general upward trend in the number of people. For long periods, civilizations stay ahead of famine, and excessive consumption of material goods occupies their thoughts, while around the Mediterranean the forests of cedar vanished except for pitiful vestiges in a few cemeteries. Affluence, at least for a few, existed for a while in such civilizations as Greece and Rome, in Israel under Solomon, in Babylonia under Nebuchadnezzar, and under the Pharoahs in Egypt.

But how does nature's agriculture differ from the human agricultural system? Wind blows. Water falls and runs. Humans give the water advice with systems of ditches, but they are soon neglected and fill in. Wind and water in nature's systems have contributed to the health of the placenta. Wind and water in the human agricultural enterprise remove the land and eventually cover the handiwork of civilization. But why?

Let us find a patch in the North American Great Plains, where vegetation has been destroyed. We can't tell whether it was ruined by drought and wind, or by a flood or the trampling of millions of buffalo. Although this happened before White people invaded the continent, it looks very much like an area laid bare by some agriculturists on the other side of the planet. In the area we presume was trampled by buffalo, succulent growth appeared in the following year, hale and hearty-looking. Though there are many kinds, patches of one species are not uncommon in this first stage of succession.

The field on the other side contains only annuals and looks like the area destroyed by nature: the vegetation is similar, soil run-off about the same. In the second year following disturbance, nature grows a different ensemble of plants from the first year's, and more perennials move in. In the human fields, the crops are the same crops as the first year's or similar. *All are still annuals!* Soil erosion is minimal in nature's field but remains the same on *Homo sapiens'*, where by the fifth year, the placenta does not heal, but continues to erode.

A profound truth has escaped us. Soil is a placenta or matrix, a living organism which is larger than the life it supports, a tough elastic membrane which has given rise to many life forms and has watched the thousands of species from their first experiments at survival, many of them through millenia-long roaring successes and even dominion before their decline and demise. But it is itself now dying. It is a death that is utterly senseless, and portends our own. In nature the wounded placenta heals through plant succession; enterprising species cover wounds quickly.

The human agricultural enterprise and all of civilization has depended upon fighting that succession. The human purpose has so dominated our thinking that those in high places are out of touch. David Brower, in his lectures, sometimes tells this story: "An eminent forester, looking at a natural process in a redwood forest, once said, 'Nature never does anything right,' and thereby substituted hubris for humus. He was wrong of course."

He wasn't only wrong; he continued to nurture the very germ of the split between the human and nature from which we humans are ultimately inseparable.

2
The Failure of Success

> Early the next morning the Spaniard went ashore with several of his men. "When we reached the land," he writes, "our first act was to fall down on our knees and render thanks to God and the Blessed Virgin without Whose intervention we had all been dead men." Their next act was to "take possession" of the land in the name of the King of Spain, and to ground the flag. As we read to-day of this solemn ceremony, its pathos and puny arrogance touch us with pity. For what else can we feel for this handful of greedy adventurers "taking possession" of the immortal wilderness in the name of another puny fellow four thousand miles away, who had never seen or heard of the place and could never have understood it any better than these men? For the earth is never "taken possession of": it possesses.
>
> —Thomas Wolfe[1]

> We was always taught (I was, I know) that a pioneer, by golly, was a hero. No question, you know? But did you ever think, really, folks, that a pioneer was nothing more than a guy that cut down a tree! And he plowed up land that probably should have been left to grass. You know, that's why we got our problems, down there in the Southwest, in my home. You know, they call it the Dust Bowl now. You see, folks, what we're learning today now is that you can rob from nature just the same way that you can rob from any individual. It ain't just robbin' from nature. It's robbin' from future generations.
>
> —Will Rogers[2]

Some things make no sense. Consider these paradoxes of our time. There is less soil on our fields each year, but there is more total production from the fields. The soil becomes increasingly poisoned from farm chemicals and salts from irrigation, and still there is more production. A million acres a year are lost to urbanization, and production climbs. There is a continual decline in the variety of germ plasm, and therefore our major crops are increasingly vulnerable to pests and diseases, yet crop production climbs. There is less water for irrigation in our aquifers, and yet more total water is being pumped, contributing to an increase in production.

What kind of a land is this that allows such shocking paradoxes? Understanding comes only when we consider the entire North American continent that Europeans confronted when they stepped ashore. A. H. Williams once wrote:

> The physical configuration of a continent, as in the case of an animal, is determined largely by the skeleton. Here we had three large pieces of bony structure: In the West there were the western Himalayas, two high mountain ranges; within those ranges, a high, dry plateau extending all the way from Alaska down the North American continent through to South America. Over in the East we had the lower, smoother Appalachian chain, and up at Nova Scotia, swinging around in an arc, up around the Hudson Bay, we had the Laurentian Mountains.[3]

Between these bony structures, these mountain barriers, was the largest expanse of temperate-zone fertile land in the world. A major part of it had been touched by the glaciers. This part also happened to receive plenty of rainfall to solar irrigate the adaptable crops from the old countries, as well as the crops already here, particularly tobacco, corn, squash and beans. We couldn't fail, really, for it was a land rich in minerals—coal, iron, copper, petroleum—and, as Williams noted, "easily approachable from the European side."

This was our heritage, and forty years ago it would have made sense to attribute a good part of that heritage exclusively to the uplift of those bony structures which gave our continent its definition.

Here is the source of nutrients which became loosened and ran down to the valleys or were blown sooner or later to form the fertile hills of the American countryside. This was the basis of our agriculture and culture. Changes in agriculture in less than a half-century now have us basing our way of agriculture and our way of life on petroleum and coal—the sunlight captured and stored even before most of the uplifting activity that gave us our mountains.

In only four short decades, as the soil was rapidly giving out, the American farmer shifted from soil to oil.

In the twenty years from 1949 to 1969, American agriculture increased its output fifty percent. During the same period of time, it was withholding from production a net land area of fifteen percent, totaling fifty-eight million acres.[4] Land was also being taken for the construction of a massive automobile transportation system and for urbanization. However, agricultural yields increased six percent per year, more than offsetting land loss. We grow increasingly more food on fewer acres and, in 1978, exported over twenty-seven billion dollars[5] worth of farm products a year on a planet where people are hungry and starve by the millions. There is a strong temptation for us to believe we must be doing something right.

Unfortunately, our successes are measured on discount economics.

At the time when we had seriously depleted the life-giving capability of the land, American agriculture began the heavy fossil-fuel chemotherapy which has given us all a false sense of the health of the agricultural system, even as it is being poisoned and further depleted. At the moment, we are poisoning the North American continent with pesticides and fertilizers, salting millions of acres through irrigation, and promoting erosion, through our methods of cultivating, of tens of millions of acres of top cropland.

If we select corn as the symbol of our agriculture (it is our top carbohydrate producer), we can say without any exaggeration that corn, as a technological product, has reduced more options for future generations than the automobile.

Less Soil—More Production. The opening paragraph of the "digest" in the General Accounting Office's report to Congress in February, 1977, reads as follows:

> Estimates of soil losses from 283 farms GAO visited on a random basis in the Great Plains, Corn Belt and Pacific Northwest indicate that topsoil losses are threatening continued crop productivity.

The contents of the study, as one would guess, are equally alarming, but the public's ignorance of the GAO report is even more so. Discounting the future in the "hundreds of years," the Soil Conservation Service determined that the "acceptable" annual loss for deep soils is about five tons per acre; for shallow soils, one ton. According to the study, eighty-four percent of these typical farms were losing *over* five tons an acre on cropland.

Let us look at these numbers another way. Soil one inch thick over one acre will weigh 150 tons.[6] So, if we take the Soil Conservation Service estimate before 1979 of nine tons per acre per year nationwide, we lose an average of one-sixteenth inch of soil per acre. In sixteen years we lose one inch of soil. In 100 years, we lose six inches!

Estimates of how long it takes for one inch of topsoil to be created, under natural conditions, range from 300 to 1,000 years.[7,8,9] In agricultural systems, when fertilizers replace plant nutrients and when organic matter is allowed to accumulate, one inch of topsoil will be rejuvenated in 30 years.[10] Except on Amish-like farms, the fertilizers that are made available are made from fossil fuel.

We have had the Soil Conservation Service some forty years now, with expenditures for their programs running into the billions. One would think that if soils aren't in better shape today, at least they would be holding even. But what do the numbers show? Various soil loss estimates were really bleak until recently. The SCS had the lowest going: 9 tons/acre/year. In 1979, this was reduced to 5 tons/acre. We should be cautious about this one-year estimate because it doesn't reflect an average over several years. Also, soil loss is so important, ordinary prudence causes us to pay close attention to estimates made by organizations other than SCS. An Iowa State University research study in 1972 estimated that the United States was losing over four billion tons of soil each year through water erosion. Compare this figure to 1934, when it had been estimated that we were losing three billion tons. Since these are just numbers, the writers of the report translated them into freight cars, loaded to their rated capacity. Such a train would be about 633,000 miles long, stretching to the moon and back and towards the moon again, or, to stay closer to home, the train would encircle the planet twenty-four times. Remember that this is the loss from the United States alone.

The total soil losses and per acre losses reflect the fact that less land was in production in 1972 than in 1934. According to this Iowa State study, in thirty-eight years the gross per-year soil loss increased from 3 to 4 billion tons per year but the *average* per acre loss jumped from 8 to 12 tons. The only gains, presumably, were in the already fat deltas far away.

In his second message on the environment, from the detailed fact sheet for new initiatives released August 2, 1979, President Carter acknowledged that before 1935, the U.S. "had well over 600 million acres of actual or potential cropland." Since then, however, "approximately 100 million acres of potential cropland" have been "effectively destroyed." On an additional million acres, half the top-soil has disappeared.

The Soil Conservation Service, looking at the state of Iowa for the month of June only in 1974, reported losses of forty and fifty tons per acre on *ordinary* land, and as much as 200 tons on unprotected slopes. Just a final clincher in this regard seems in order. A January 1975 report from the Council for Agricultural Science and Technology apparently agrees with the Iowa State report, for it suggests that we are less effective at controlling erosion today than we were even fifteen years ago.

It is well known that current dollar costs indicate little about the long-term costs to civilization. An economics which discounts the future is inherent in the environmental problem. Even so, the dollar costs to remove this "ecological capital" out of ditches, streams, rivers, reservoirs, harbors, lakes and ponds is substantial. The Army Corps of Engineers worries about such phenomena as conveyance capacity of rivers, hindered navigation, and damage to Corps-constructed recreational marinas.

The U.S. Department of Agriculture has estimated that plant nutrient losses owing to erosion amount to around $18 billion per year in 1979 dollars.[11] This is just to replace the nitrogen, phosphorus and one-fourth of the potassium. To these higher production costs should be added the costs for degraded water and soil quality and, in some cases, the lowering of agricultural yields. The soil microflora, which assist in the breakdown of pesticides, is decreased as soil washes away. President Carter's message on the environment concluded that wind and water erosion "cause losses of organic matter important in the retention of soil moisture, the survival of soil microflora and the chemical immobilization of toxic metals."

In forty-one years, from 1935 to 1976, our federal government has spent $20.7 billion "to conserve soil and water and to increase agricultural productivity."[11] This cost includes "funding for cost-sharing, technical assistance, resource management, loans, research, and education." A sobering conclusion of the 1977 GAO report noted that the "recipients often did not fully implement suggested or required programs" and spent the money for production-oriented efforts. Follow-up by federal officials to encourage real soil conservation programs was practically nil.

One wonders how a great technological civilization such as ours would allow such atrocities to occur to the land and to the future. One is reminded of the Sumarians of Mesopotamia who invented the wheel six thousand years ago and whose mathematics we still employ in our clocks, which divide time into units of sixty. In one sense they were the best of peoples at their time, but in a resource conservation sense, they were the worst. Little grows over much of their old lands now. How about the enlightened Moses who led the Israelites to the Promised Land? It was "a good land, a land of brooks of water, of fountains and depths that spring out of the valleys and hills; a land of wheat and barley and vines and fig trees and pomegranates, a land of olive oil and honey; a land wherein thou shalt eat bread without scarceness; thou shalt not lack anything in it." Three thousand years later, bedrock is exposed throughout the uplands with the soil now in the narrow valleys or at the sea. Civilizations that are fountains of knowledge do not necessarily exercise wisdom.

One wonders whether a whimsical remark made many years ago by a U.S.D.A. soil scientist has taken on a profound truth. The scientist remarked that the ultimate goal of plant breeders and agronomists is "to grow forty bushels per acre of wheat on dry bedrock."

Erosion Control Through Appropriate Methods. When we say we are losing so many tons of soil per acre each year (depending on whose study we accept), we have collapsed together many numbers which prevent us from seeing the severity of damage in any one place. To get specific we must consider soil type and depth, slope, including length of slope, how much organic matter is present, cultivation practices, what crops are grown and in what kind of a rotation schedule, and finally, the intensity and duration of wind and rain. The average corn or cotton farmer who never rotates will experience a loss of around twenty tons per acre.[12] If the same land is in forest, there will be a loss of 0.01 to 0.002 tons per acre.[7,13] Under normal agricultural conditions, top soil can be formed at a rate of 1.5 tons per acre per year.[6]

Contour planting requires a little over five percent more time and energy, but the payoff can be significant, depending on the type of weather.[6] In the eastern United States where rains are gentle, very little soil may be lost. In the Great Plains and Midwest where the thunder storms are dramatic, soil loss will likely be much beyond the replacement level, even with contour planting.

It would seem that farmers who rotate their crops lose less soil. A typical Georgia cotton farm with a seven percent slope will lose over twenty tons per acre. But when cotton is grown in rotation, the farm will lose around six tons per acre—still much greater than the rate of replacement.[12] Land devoted exclusively to corn in Missouri will lose nearly twenty tons per acre, but similar land in rotation with wheat and clover has lost only 2.7 tons per acre.[14]

But what does all this really mean? The soil loss savings owing strictly to the time factor in crop rotation is no savings at all if a fixed amount of corn is produced. In fact, it may be a loss. It seems more sensible to consider the tons of soil lost per bushel of corn produced, period, regardless of what the intermediate forage crops are. Let us assume, for a moment, that we have to raise 100 bushels per year to meet our needs and that we have three acres to raise it on. One is a flat acre which can produce, with appropriate chemotherapy, 100 bu/year. The other two acres are steep acres which can produce only 50 bu/year. The flat acre loses 10 tons of soil per acre per year; the steep loses 20 tons. Do we farm the flat area each year and give up .1 ton of soil per bushel produced? Do we farm the hills each year and give up .4 tons/bushel, or do we alternate each with a fallow year for an average of .25 tons/bushel? This is not a hypothetical question—this is a real decision the country will have to make.[15]

Livestock manure is of tremendous value in holding the soil. Its spongy nature absorbs the blows of the rain and the water itself. Sixteen tons of manure on a nine percent slope in Iowa reduced the soil loss in one year by over seventeen tons.[7] It is clear that organic matter is meant to be left in the field and in no way regarded as waste, a point we will come back to.

Many farmers plant crops which germinate and cover the ground fast, and they leave them in the field during the time the main crop is not being grown. This reduces nutrient loss and increases the organic matter. Some farmers will interseed a legume in their cereal crop and thus take advantage of its nitrogen-fixing qualities during growth; it then provides a ground cover when the main crop is harvested.

The latest success story in controlling soil loss is the minimum- or, in some cases, no-tillage method. The farmer does not plow, disk or harrow before seeding; instead he opens a small furrow in the already green vegetation and drops the seed. An herbicide is applied to reduce the competition, but the soil is held in place by the roots which were there in the first place. In Nebraska, soil loss was cut to one third in some tests (10.7 vs 3.4 tons/acre).[16] But in Ohio the loss of soil from no-tillage corn averaged less than 1/100th that from corn grown in a conventional manner.[17] This method may be of short-run success since we do not yet know the consequences of chemical poisoning. When we begin to add them up after years of this method, we might find that the health effects are very serious, and that minimum tillage is a bad bargain.

American Farm Policy: Simply Discount the Future. We have seen that the average annual sediment losses in the last forty years have been estimated to have increased a full twenty-five percent, from three billion to over four billion tons. We are aware that this is directly correlated with our successful attempts to increase production. Some of these erosion levels, particularly on the Great Plains, have greatly increased in the last twenty years. U.S. government policies in the early 70's under agriculture secretary Earl Butz encouraged fence-row-to-fence-row farming. Thousands of miles of shelter belts were pushed out, conservation projects of long standing were plowed and planted. In western Iowa, erosion leaped twenty-two percent in this short time.

David Pimentel and his collaborators at Cornell University have noted how the short term economic effects of soil erosion can be minor in one year when we look at a crop like corn. "Assuming that there is a reduction of four bushels per acre in yield for each inch of topsoil lost," they say, "and that about twenty tons of topsoil per acre are lost annually in continuous corn production, then the annual per-acre reduction in yield (from land with an initial topsoil depth of twelve inches or less) would be about one-quarter bushel of corn (worth about $1.50). When per-acre costs of corn production are estimated at $190, the $1.50 represents an annual loss of less than one percent, a negligible amount for a single year."[6]

When short-term costs for soil conservation procedures are greater than short-term returns, there is little doubt which route most farmers will take. In the corn country of northeastern Illinois, contour planting and rotation reduced erosion to around 2.2 tons per acre compared to 18.8 tons for continuous corn cropping. Over a twenty year period, based on a five percent discount rate, the conservation practice penalized the farmer $39 per acre.[6]

The national Commission on Water Quality estimated a cost of $3.10 for conservation practices on irrigated cropland.[18] At least one other study has shown that reduction of average annual soil loss has decreased the net farm income.[19] Here is a clear example of the insensitivity of our economic system to the needs of future generations.

Finding the Bottom Line. Fertilizer spread on completely denuded rock would probably produce a few grains of wheat in a year. The inevitable cracks in the rock contain enough soil to support a few plants. After all, grass grows in cracks of concrete. In many respects the soil loss problem is like the fossil fuel problem. We will never run out of oil, but the time will come when the energy required to lift a gallon will be equal to the energy present in that gallon. Actually, economics will likely have stopped us before that time, unless we want oil for purposes other than the energy in it. We may need it for lubrication.

Similarly, at some point before we reach that bedrock when the soils are thin and poor, fossil-fuel-backed fertility will be the most energy-expensive part of our lives. The farmer will not need to go on strike to obtain fuel. After all the talk and passion has subsided, our society will survive with a minimum of heat and jet airplanes; but we will still need food every day, and we will insist upon it. Our last available fossil fuels will be spent on agriculture, perhaps to make nitrogen fertilizer to offset the consequences of soil loss.

Nevertheless, even this system will eventually come to an end, for any life support system will eventually face problems if the mines and sinks are so far apart that the contents of the sink cannot recharge the mine. In the grasslands of North America, Bison could mine the grass and Indians could mine the buffalo so long as the sinks (places that would receive the digested biotic material) continued to re-supply the mine. For such a system, words like mines and sinks are unnecessary. As long as nutrient cycling is complete, or nearly so, we can say that the buffalo and the Indian "harvest."

The ways of our modern society force us to think in such terms as mines and sinks. Livestock and grains are removed (mined) from our fields, run through the human digestive track, deposited in sewers and eventually into the sinks we call streams. This system can't last, because the mines and the sinks are too far apart.

Losses to Urbanization. Historically, people have placed cities near a food supply and city expansion is mostly over level, well-drained land with rich, deep soils. Such a settlement pattern provides the city with an ample water supply and efficient disposal system via the nearby river. Right now in the U.S., around 2,000 acres of prime farm land are gobbled up by urban sprawl each day. This amounts to a million acres a year of the world's best and most level land.[6] Secretary of Agriculture Bob Bergland recently illustrated how dramatic this loss has been: "In my lifetime, we've paved over the equivalent of all the cropland in Ohio. Before this century is out, we will pave over an area the size of Indiana."[5]

Gus Speth, Chairman of the Council on Environmental Quality, has said, "Farmers, ranchers and environmentalists are allies now in their concern for the good stewardship of the nation's agricultural lands. This partnership is an immensely important one. Soil is the raw material of agriculture. We can pave it or we can save it, knowing whatever choice we make will profoundly influence the lives of unborn generations."[5]

Chemotherapy on the Land

> They claim this Mother Earth of ours for their own and fence their neighbors away from them. They degrade the landscape with their buildings and their waste. They compel the natural earth to produce excessively and when it fails, they force it to take medicine to produce more. This is evil.
>
> —Sitting Bull, 1877

Barry Commoner mentions in *The Closing Circle* that between 1949 and 1968 the "annual use of fertilizer nitrogen increased by 648 per cent." In about the same period (1950-1967) pesticide use in the U.S. increased by 168 per cent *for each unit of agricultural production*. At the end of World War II, Illinois farms averaged 50 bushels of corn per acre. Twenty years later these same farms produced 95 bushels per acre, almost a doubling in yield. But in 1945 those fields "consumed" only 10,000 tons of chemical fertilizer compared to 400,000 tons twenty years later. The near doubling in production required a 40-fold increase in the application of the energy-intensive fertilizer product. In recent years, the feedstock for this fertilizer has amounted to well over one-fifth of the interruptable supply of natural gas.[20]

These are staggering numbers, but with the most simple kind of cost accounting, such widespread and intensive treatment of the land has been a bargain. What most of us have missed, however, is that agriculture as an industry is now the leading polluter of streams and ground water. From farm sale talk, I hear rumors of farmers in my area who, upon renting a piece of ground to grow livestock, are apt to seek an agreement with the landlord ahead of time which would allow them to back out of the contract in the event that the water is so loaded with nitrates that the livestock are threatened with death or illness.

One has to be careful when tuned in to a rural grapevine. Under the local gossip is a current of subtle humor and understatement that makes it difficult for an outsider to distinguish between just plain "jawing" and what is really happening to the people. I suppose no major newspaper is up to the task of sending out someone who doesn't act like a reporter, but if it could, it would find stories about the farm families' miscarriages, the spontaneous abortions of pig litters, calves and foals, the families' headaches (until they switched to bottled water) and, in one case, blindness in children.

Many water samples analyzed—or "sent in to the state," as we say—throughout the midwest and Great Plains are excessively high in nitrates. Fertilizer runoff is charged as the major culprit, but the densely populated feedlot also makes a major contribution to poisoned aquifiers and streams. Fish kills in the rural countryside are not uncommon.

Eight communities in Nebraska have recently confirmed nitrate levels in the water supplies in excess of the federal standard of ten parts per million. Ten additional communities are suspected of having too high a concentration. These communities will have to either hook up to another rural water supply district or drill new wells or they will be forced to install equipment costing $100,000 or so to remove the nitrates, and then they will face an additional annual expenditure of $20,000 to operate the equipment. The Nebraska Department of Health believes these high levels are associated with increased irrigation, especially in areas with sandy soils.[21]

About 40 million tons of commercial fertilizers are being applied to American fields each year.[22] We have long known that nutrient-rich water ecosystems experience choking plant growth and reduced oxygen levels. Some decaying algae release toxins and further add to the problem. The accumulation of this ecological debt now amounts to over a million metric tons of nitrogen in our surface and ground waters alone. To this must be added two billion tons of livestock manure, at least half of which is unregulated by any level of government.

Competition from other species for the human-grown food and fiber supply is more intense than ever. Let us say we are losing between 33 and 42 percent of all that we grow.[23] Taking the low estimate, insects get about 13%, pathogens 12%, and weeds around 8% of total production. This is in spite of all the pesticide and biochemical controls in use. Crop losses from weeds have been slightly reduced since 1942; losses from pathogens have increased 1½%, and insect losses have doubled at a time when insecticide usage has increased ten fold.[24] This 33% total loss in production amounts to around $35 billion worth of crops, which, if it could be saved, would have been enough to pay for all the oil imported in 1976.[24] As substantial as this is, it wouldn't come close to the more than $83 billion that foreign oil cost in 1980.

Others have complained of these shortcomings in American agriculture, and have accurately pointed their fingers at the causes. Few of the causes are defensible. While we push for production, we plant more crop varieties that are susceptible to pests. Crop rotations and crop diversity have declined, especially with crops like corn. We are now less sanitary in our fields and orchards than when we destroyed the bad apples or infested twigs and crop residues. We also grow some crops, like potatoes and broccoli, in climatic regions where they are more susceptible to insects.

As always, the insects and pathogens have evolved resistance. In the course of it all, we have destroyed many of their natural enemies, thus increasing the need for more chemical treatments. Furthermore, chemical availability has allowed the breeders to alter the plants' physiology, often unwittingly. If a plant has a physiological resistance to a particular pest, the genes that control that resistance do not have to be retained in a breeding program if a chemical application is available. In such a manner, several of our major crops have rather recently evolved a "weakened" condition. Corn is a good example of such a plant. Only a few years ago, through some fortunate weather, the American corn crop was spared widespread blight. A few years before, a male sterile factor had been discovered in corn and was quickly incorporated by the breeders. This genetically controlled factor, it turned out, was present in the cytoplasm of the cell, that part of the cell outside the nucleus. (The nucleus is where most of the genetic material is located and arranged on the chromosomes.) This "cytoplasmic gene" for male sterility was valuable to the seed companies, for the plant would emasculate itself and allow the plant breeders to bypass the expensive hand labor of physically removing the tassel or pollen-bearing portion of the plant. Unfortunately, the corn wasn't resistant to a new strain of blight; it could have taken most of the crop. By the next year the seed companies had responded with seeds carrying genes for resistance to the blight.

Professor Pimentel's group has tried to determine what our crop losses would be if all pesticides were withdrawn and readily available non-chemical methods were used. Their preliminary findings suggest that our crop losses would probably increase from 33 to 42 percent. But that is if we take all crops into consideration. If we were to subtract off the non-food crops such as cotton, tobacco, hay and pasture, so that we are looking at the loss of food crops only when grown without pesticides, we have another story. The increase in loss goes up nine to eleven percent.[24]

Based on their studies and estimates, Pimentel and his collaborators have concluded that there would be no serious food shortage in the U.S. if crops were not treated with pesticides. Our total variety of food would steeply decline, but we'd still have plenty to eat. It is just that we would have fewer fruits and vegetables such as apples, peaches, plums, onions, tomatoes and peanuts.[24]

Discounting health entirely, control through pesticides yields $4 for every one dollar spent. Although only 200 people die every year from poisonings, the EPA has estimated that some 14,000 individuals are non-fatally poisoned by pesticides each year, and 6,000 are injured enough to be hospitalized.[24]

The use of pesticides is another example of using a sledge hammer to kill an ant. About 1% of the pesticide hits the target pests.[24] Often only a quarter to a half reaches the crop area, especially from aerial spraying. and 65% of all agricultural insecticide is applied by plane.[24] Herbicide application has become a bit more sensible in recent years, with machinery designed for the spray to hit the weeds that stand above the crop only. Nevertheless, as we have seen, overall it is a fossil-fuel-based system of control, for a full 80% of the one billion pounds spread annually comes out of the oil wells, even though it amounts to only 0.04% of U.S. oil consumption in 1980.

The late Professor Robert van den Bosch, a research entomologist of international renown at University of California, Berkeley, and author of *The Pesticide Conspiracy*, summarized the problem dramatically shortly before his death:

> Pesticides are a hazard to human health and to the general environment. It should be obvious to any perceptive person that if we want to reduce the hazard, we must junk the chemical strategy. However, in contemplating such a move, it must first be fully understood that there is no simplistic alternative.
>
> We continue to be fascinated by the prospect of a simple miracle, and repeatedly reach out for it. We must stop this nonsense and instead concentrate our energy, resources and brainpower on the only workable pest management strategy available to us—integrated pest management (IPM)—which is in fact nothing other than scientific pest control.
>
> IPM is an ecologically based pest control strategy which relies heavily upon natural mortality factors such as biological control and pest inhibiting climatic conditions, and which employs control tactics that preserve or augment these factors or the mortality they cause. IPM uses pesticides, but only when special circumstances indicate a need. And even here the appropriate kind of material is used along with proper dosage, placement and timing of application.
>
> All IPM programs of which I am aware are equally effective to—and no more costly than—the conventional chemical control program they have replaced, and many have been more effective and less costly. In other words, IPM does not sacrifice pest control efficiency for environmental quality, as some detractors infer.[25]

There needs to be financial support on this end of the problem. The pesticide companies certainly have plenty, as any issue of the *Farm Journal* will show. Organic methods have little in the way of money to promote them. Nevertheless, we can regulate the poison peddlers who hustle their products, and, as Professor van den Bosch says,

> . . . vastly upgrade the qualification of licensed pest control advisers, tighten pesticide registration protocols, insist on the evolution of safe-selective pesticides (which can best function and prosper economically under IPM), take pesticide use decision making away from all unqualified persons, support research to develop IPM programs and implement these programs as rapidly as good science permits.

Humans are sloppy creatures, but they seldom err on purpose. For each action there is some degree of forgiveness, some level of tolerable sloppiness. Training and licensing of pesticide handlers will not completely eliminate contamination of the landscape, and we must recognize that most of the total load of poisons in the environment will be there due to error and not intent.

Irrigation. A water crisis "is a clear threat by the turn of the century" says the U.S. Water Resources Council, and agriculture must shoulder most of the blame, for over 80% of our water is consumed on farms and ranches.[26] Given the current level of salt pollution, the SCS has estimated a cost of $3.2 billion to keep the problem at a manageable level. To control current pesticide and fertilizer pollution will take an additional $1.8 billion. To move the already deposited silt and sediment from our soils and away from our water will cost $4.7 billion. Not all of these water problems are caused by irrigation, but irrigation is a heavy contributor.

Some 44 million of the 400 million acres of American cropland is irrigated, amounting to 12% of the total.[26] The two major problems associated with irrigation agriculture may or may not occur together. On the one hand, there is the problem of salt accumulating in fields, a common problem when water is pulled from such western rivers as the Colorado. The second major problem is aquifer depletion, where drawdown from pumping occurs at a rate faster than recharge. This is the case along the huge High Plains Ogallala aquifer, which runs from Texas to Nebraska.

Seven hundred thousand tons of salt are added to the Colorado each year from the Grand Valley of western Colorado alone. The consequence is an economic impact of more than $50 million each year in reduced crop yield and corroding of equipment.[26] This cost could easily double in the next 20 years, even at a fixed dollar cost. Salinity at the headwaters of the Colorado system is less than fifty milligrams per liter. At the Imperial Dam near Mexico, salinity has increased eighteen-fold to 900 milligrams per liter of water.[26] The concentration is expected to reach 1160 milligrams/liter by the year 2000. The economic losses for each one milligram per liter addition is over $200,000.[26]

In the Great Plains states, where water mining is most intensive, many areas have pumps pulling water ten times faster than it can be replaced. Much of this aquifer mining has used cheap fossil fuel. Even with electric motors to run the pumps, the energy cost has been low. Let us assume a cost of 5¢ per kilowatt-hour of power and that the electricity represents half the pumping cost, capital and maintenance making up the rest. With a 75% efficient pump motor, an irrigator would have to spend $25 to lift an acre-foot of water 188 feet. This is too expensive for agriculture. Several irrigation projects completed in California have water for sale at this price and no takers.

For most irrigation projects, however, the water user has little economic reason to limit the amount diverted to meet his demands. Figures for 1972 in California,[27] an almost exclusively irrigation-dependent state, reveal that water is an incredible bargain, as energy used to be. For example, over one-and-a-third million acres of alfalfa received an average of five feet of water over the entire area. This is about 6.73 million acre-feet. The value to the producers from this acreage was around $252 million. So for each dollar's worth of alfalfa hay, 9,080 gallons of water were used. Slightly fewer than a million acres of California cotton were irrigated with 3.875 million acre-feet, or slightly over four feet of water spread over each acre. The gallons of water per dollar income for the producer amounted to 4,306. As underground aquifers dry up, cotton and alfalfa hay will likely have to move to a more humid climate, and areas with little rain will have to revert to dryland farming.

A third California crop, 545,000 acres of grapes, consumed 1.903 million acre-feet, or 3.5 feet of water per acre in 1972.

California grape growers received $368 million. This amounted to 1,758 gallons per dollar. These three crops account for about a third of both acreage and water used by the more than 200 commercial crops grown in California.[27]

All this is applied water and it is hard to say what fraction is consumed, but more than half. Some is recycled as return flows.

The American people have been through an era of cheap energy and have assembled an economic system based on those low costs. We are just beginning to see the convulsions that are likely to happen now that the cheap-energy era has come to an end. The water problem may be ten years down the road, but as with energy, much of our food-producing system is based on very low water costs. We have dammed rivers and salted valleys. We have mined aquifers and have built an economics around cheap water.

Narrowing the Genetic Base. Plant breeding programs have played a significant role in setting production records through the genetic improvement of our major crops. But here is another area where success is tied to a time bomb by a short fuse.

Short-run gains have come at the expense of increased vulnerability through a narrowing of the genetic base of many major crops. How it happens can be illustrated by one small example. Resistance to a particular pathogen required genes that dictate that a certain immunological defense mechanism is in force to keep the disease under control. The maintenance of that defense mechanism requires energy, which the plant collects from the sun. When the defense mechanism is eliminated from the plant, more of the sun's energy can be allocated toward harvestable production for humans.

For purposes of illustration, I have mentioned resistance to disease. The same holds for resistance to insects and even drought. Again, insecticides and irrigation solve many problems for plant breeders and allow them to pay closer attention to the bottom line.

Some of this genetic selection happens unwittingly, because the seed companies are part of our fossil-fuel-based agriculture. Plant breeders are afforded the luxury of ignoring many genes for resistance when they make their crosses. The upshot is that our major crops have experienced a steady narrowing of their genetic base. This is called genetic truncation.[28]

The situation is bad enough in the U.S., but many underdeveloped countries, seeking to increase their food supply, have encouraged American-style agricultural scientists to come in and show them how. The "improved varieties" of wheat, rice and corn have driven many of the old, local, reliable varieties from the fields in many places around the globe. The irony of it all is that much of the genetic resource base upon which the new variety is built is being displaced.

Water mining, land salting, pesticide accumulation, fertilizer application and genetic vulnerability are of the same cloth. Any time short-run efficiency, affordable because of huge energy injections, is the major consideration, the temptation to reduce biological variety is compelling. In the distant past, our ancestors manipulated the genetic program every time they planted and harvested. In that sense nothing is new. But most of the pre-fossil-fuel era activities over the range of a given species *promoted* genetic variation. One area had a strain well-adapted to local conditions and another area had a different strain. On the whole then, variety was being generated, even though the other parts of our agricultural enterprise were reducing options for the future. Now the last piece, genetic wastage, seems to have fallen in place with the rest of our eroding options, an undeniable example of failure due to success.

The $5,000 Flat Tire

The problems of farmers which force mistreatment of land are poorly understood by urban dwellers. Many farmers have little choice and move reluctantly into a large-scale, capital-intensive type of farming. Hundreds of farmers are frustrated as they try to extricate themselves from the trap they are in. They have every reason to be anxiety-ridden, especially when they reflect that some of their former neighbors have been forced off the farm and that they could soon follow if something isn't done about those large notes at the bank that will come due in a few weeks. Admittedly, part of the problem in many cases is of their own doing. In the past, the big team and large harness were prestige laden. Now that prestige belongs to the large and new tractor.

I was reminded of the expensive problems of farmers in the first days of summer in 1975. I had gone home to help my brother, a Kansas grain farmer, during wheat harvest. Right at the busiest time, when all motor-driven equipment was essential for the variety of work which needed doing, a rear tire on his 1972 John Deere tractor went flat. These are huge tires, filled with a liquid to add weight and traction. Special equipment is required, not only to remove and replace the heavy tire, but to pump the liquid out before repair and pump it back into the tube later. Farm machinery and tire dealers usually have the necessary equipment on a service truck and are ready to assist farmers when tire trouble develops.

Someone discovered the flat in the late afternoon, and a phone call brought the service man out shortly after sunrise the next morning. He took a quick look at the tire and cheerfully went through the motions of fixing it. We had breakfast and my brother outlined the day's work, which included using that tractor. Unfortunately, the service man put the tube in backwards, and a second flat developed and ruined both the tire and tube. No tire dealer in the area had a replacement of that particular size and after almost a week with the tractor out of production, my brother "dualed" the rear wheels. That is, he bought four smaller-sized tires to replace the two. The cost for this modification was around $1,200, a cost he would not have had if another tire had been available.

If we stopped right there, the story would be bad enough. The largest cost of that flat was incurred from the week the tractor was out of production. He has estimated that this lost time cost him $4,000–$5,000. It happened during early harvest, which is also planting time for such row crops as sorghum and soybeans. Furthermore, my brother could not deduct that loss from his income tax, since he hadn't realized the income, so his loss was substantial.

The lost time cost so much because farming is not like ordinary production industry. If industry loses a week, it has only experienced a reduced earning on the capital investment. After the lost week, production can continue as before. But planting and harvest are seasonal and weed control must be counted in. These activities require the right equipment being in the right places at the right time. Successful farmers have paid close attention to these details.

Let us accept then that this one flat did cost $5,200 ($4,000 for lost field time and $1,200 for being forced to "dual" the tractor) and that his 30 bushel/acre wheat was selling for $3 a bushel at the local grain elevator. The *gross sale* on nearly 58 acres would be necessary to pay for one equipment problem. However, the production costs on that $3-per-bushel wheat amounts to about $2 a bushel and here I am assuming no cost for renting the ground, no payments for buying the ground, no taxes on the land. With a profit of $1 per bushel, it would take 175–200 acres of wheat to pay for that one flat tire.

It is easy to see how a combination of land prices and operating overhead have forced so many farmers out of rotation-type agriculture. One of the reasons that flat cost my brother so much is that he has a diversified operation. When farmers are struggling to make a profit, who can blame them for promoting high plant densities of one crop?

In the late fall of 1979, I was flying from Sioux Falls, South Dakota, to Kansas City with a brief stop in Sioux City, Iowa. At Sioux City, a 70-year-old northwest Iowa farm couple, husband and wife, boarded and sat facing me.

A month or so earlier I had seen on the CBS news that a rain and ice storm had come through the region and caught farmers with at least 25 percent of their corn still in the field. I could readily see this fact of life from the air, but to make conversation I asked the couple how things were going. The man responded that "things was rough." He went on to describe an almost unbelievable array of local problems. The weather had caught several farmers with their harvesting equipment in the field. One farmer's equipment became so buried in the mud that he hired a backhoe and operator to dig it out. The result was a hole so large the neighbor next had to rent a bulldozer to fill it in. Another neighbor's equipment became so mired in that he hired a heavy-duty helicopter to lift it out. The cost—$4,000. A third neighbor had stuck his corn combine so completely in the mud that when they used the powerful equipment to pull it out, they literally pulled the $80,000 harvester in two. The owner went home and shot himself.

To an outsider, these farmers may appear stupid for getting stuck in the first place and doubly stupid for being so frantic to get it out. We should be careful in our judgment, for the problem is associated with a psychology that comes with being over-capitalized in the farming operation. That harvester cost as much as a house. They don't want that equipment sitting helpless in the field. (Remember the largest cost in the $5,200 flat.) They want their equipment ready, should conditions change and allow them to finish the harvest.

There is a not unusual twist to the encounter with this farmer. I had heard nearly all the bad things first. But before the plane landed in Kansas City, the farmer told me, "Ya know, I'm a millionaire. I found out last year I am worth one million seven hundred thousand." The idea made him feel somewhat pleased for he continued, "Pretty good, cause me and Mom here only had $400 and a pick-up truck when we started farming in 1940. But I tell ya, we bought land with our money. Of course we inherited some land, but we didn't go payin' International Harvester alot of money. We bought land!"

Though he was pleased with the idea of being a millionaire, he knew and I knew that his net worth of $1.7 million was a kind of joke. Inflated land prices made up most of it, and he wasn't interested in selling the hearth. They intended to pass the land on to their daughter; the inheritance tax alone may force her to sell the farm.

Simple financial accounting does not explain how this farmer could believe "things is rough" when he is worth so much. Maybe he was thinking of his neighbors' problems, realizing they had notes at the bank and that the crop in the field, still on the stalk, was the profit for an entire year.

Not long ago I talked with a farmer who explained that when he came home from World War II and started farming, he had no debts and his net worth was thirty to forty thousand. Some thirty years later he owes a quarter of a million but his assets are worth nearly three quarters of a million. He feels up against it, for in order to realize his net worth, he would have to sell out and remove his opportunity for finishing his last years doing the things he has loved, farming and feeding cattle.

I can't help but feel that the financial paradox of hundreds of thousands of farmers is but a mirror image of the physical paradoxes mentioned at the beginning of this chapter. We are talking about an increasingly powerless group of people, thousands of whom have a net worth measured in the millions, but who live in the shadow of bankruptcy and the loss of their farm. One could anticipate that the proposed strike of the American

Agriculture Movement in 1978–79 was doomed from the word go. These frustrated and over-capitalized farmers were proposing to do something that has never been done in the history of the labor movement, *withhold their capital* as well as their labor. The strike didn't come off and the problems remain.

I don't think we can regard the suicide of the Iowa corn farmer simply as the act of an unstable man. It is time for Americans to explore the reasons why his most extreme measure and the $4,000 rental paid by his neighbor for a heavy-duty helicopter are perceived as necessary in the first place. And though it sounds preachy, I think that none of us should quit until we satisfy ourselves as to why it is right that a flat tractor tire for a week on a grain farm should cost the farmer over $5,000 in real cash.

The realities we call paradoxes, be they financial or physical, give us pause because they force us to think of the reliability of the bottom layer of our food system—production itself. Later we can worry about the middlé businesses which process, package, transport and distribute our food, even though they have a heavy hand in much of the tyranny the farmer experiences daily.

References and Notes

1. Thomas Wolfe, 1947. "The Men of Old Catawba," in *Thomas Wolfe Short Stories*. Penguin Books, Inc. New York, pp. 5–6.
2. Taken from the recording Will Rogers' U.S.A., James Whitmore as Will Rogers, Columbia SG 30546.
3. Alfred H. Williams, 1944. "Reared High on Waste; An analysis of the American standard of living," *The Land* III:3; p. 235.
4. See Statistical Abstracts from 1949–1969 and the U.S.D.A. Yearbook of Agriculture over the same period.
5. Shirley Foster Fields, 1979. "Where Have The Farm Lands Gone?" National Agricultural Lands Study, New Executive Office Building, 722 Jackson Place, N.W., Washington, D.C.
6. David Pimentel et. al. Oct. 1976. "Land Degradation: Effects on food and energy resources," *Science* 194: 149–155.
7. H.H. Bennett, 1939. *Soil Conservation,* McGraw Hill, New York.
8. A.F. Gustafson, 1937. *Conservation of the Soil,* McGraw-Hill, New York.
9. O. Olivers, 1971. *Natural Resource Conservation: An ecological approach,* Macmillan, New York.
10. N. Hudson, 1971. *Soil Conservation,* Cornell Univ. Press, Ithaca, N.Y.
11. President Carter's Second Message on the Environment. Detailed fact sheet for new initiatives, page 24. August 2, 1979.
12. J.R. Carreker and A.P. Barnett, 1949. *Agric. Eng.* p. 30.
13. F.H. Bermann, G.E. Likens, T.G. Siccama, R.S. Pierce, J.S. Eaton, 1974. *Ecol. Monogr.* 44(3) p. 255.
14. M.F. Miller, *Mo Agri. Exp. Stn. Bull. No. 366.*
15. I thank Prof. C.A. Washburn, Mech. Eng., Calif. State Univ., Sacramento for calling this to my attention and for providing the example.
16. W.C. Moldenhauer and M. Amemiya, 1967. *Iowa Farm Sci.*21(10), p. 3.
17. L.L. Harrold, in Proceedings of the No-Tillage Systems Symposium (Ohio State University, Columbus, and Chevron Chemical Co., Richmond, Calif., 1972).
18. National Commission on Water Quality. *Cost and Effectiveness of Control of Pollution from Selected Nonpoint Sources,* final report (Midwest Research Institute, Kansas City, Mo., 28 November 1975).
19. K.J. Nicol, H.C. Madsen, E.O. Heady, 1974. *J. Soil Water Conserv.* 29(5), 204.

20. Guy H. Miles, November 1974. "The Federal Role in Increasing the Productivity of the U.S. Food System." NSF-RA-N-74-271. p. 20.

21. "Don't Drink the Water!" 1979. Center for Rural Affairs publication, Walthill, Neb.

22. James Risser, Sept. 13, 1978. "Farm chemicals dilemma: High yield for low water quality." *Des Moines Register,* No. 4 in a series of 7.

23. There are numerous estimates and from different times. In 1967, H.H. Cramer ("Plant protection and world crop production. *Pflanzenschutznachrichten* 20(1):1–524) estimated the world crop losses to all pests to be around 35%. Postharvest losses have been estimated by Pimentel and others ("Energy and land constraints in food protein production." *Science* 190:754–761) to range from 10 to 20%. In 1965, the USDA (*Losses in Agriculture*. Agr. Handbook No. 291, Agr. Res. Serv. U.S. Government Printing Office, Washington, D.C.) estimated the U.S. loss to be 9%. Pimentel, et. al. have estimated a 40 to 48% loss world wide when preharvest is combined with post-harvest (See "Pesticides, Insects in Foods, and Cosmetic Standards," *BioScience,* Vol. 27:3, pp. 178–185).

24. E.H. Smith and D. Pimentel, 1978. *Pest Control Strategies,* Academic Press, pp. 55–71.

25. Robert van den Bosch, Jan/Feb 1979. "The Pesticide Treadmill," *Food Monitor,* p. 23.

26. James Risser, Sept. 14, 1978. "Intensive Irrigation in U.S. threatens water supplies." The *Des Moines Register.*

27. Wm. L. Kahrl, 1979. *California Water Atlas,* State of California, p. 84.

28. Garrison Wilkes, 1977. "The World's Crop Plant Germplasm—An endangered Resource." *Bulletin of the Atomic Scientists.* See also "Our Vanishing Genetic Resources" by J.R. Harlan, reprinted in *Food: Politics, Economics Nutrition and Research,* AAAS edited by P.H. Abelson, 1975.

3
The Failure of Prophesy

Could it be that the steady demise of the forests and soils over the last eight to ten thousand years has been so slow each year that individuals during their lifetime have been unaware of the inevitable consequences? Haven't there been any prophets to warn humankind about the erosion of the life support system? Or perhaps there have been, but the prophets were ignored because they lacked facts, or style, or passion. But the record shows just the opposite. There have been many articulate, knowledgeable spokesmen who have spoken ardently about the numerous problems of people in relationship to the land.

Early Biblical directions about land use were mostly restricted to considerations of ownership. A proper attitude is described in Leviticus 25: 23, 24, where we read that, "Land must not be sold in perpetuity, for the land belongs to Me, and to Me you are only strangers and sojourners."

Land treatment was also variously impressed upon the people of the Old Testament. Job warned that "the waters wear the stones" and "the things that grow out of the dust of the earth" are washed away, eventually destroying "the hope of man."[1] The often-recalled, egalitarian concern of Isaiah's "woe unto them that join house to house . . ." sounds like a warning on the consequences of something resembling corporate farming, or perhaps of large-scale monocultures. The arrogant King Nebuchadnezzar of Babylon once delivered a familiar-sounding state-of-the-union message: "That which nothing before had done, I did . . . A wall like a mountain that cannot be moved, I builded . . . great canals I dug and lined them with burnt brick laid in bitumen and brought abundantly waters to all the people . . . I paved the streets of Babylon with stone from the mountains . . . magnificent palaces and temples I have built . . . Huge cedars from Mount Lebanon I cut down . . . with radiant gold I overlaid them and with jewels I adorned them."

While all this human handiwork was dazzling the great king and probably most of the subjects, the Hebrew prophets of the time warned that such excesses of the kingdom would turn it into "a desolation, a dry land, and a wilderness, a land wherein no man dwelleth . . ."

The Greeks were aware of the human impact upon the land. Plato called attention to the "mountains in Attica which can now keep nothing but bees, but which were clothed, not so very long ago, with . . . timber suitable for roofing the very large buildings . . . The annual supply of rainfall was not lost, as it is at present, through being allowed to flow over the denuded surface to the sea. . . ."[1]

Later, in the early Christian era, the Roman Tertullian wrote that "all places are now accessible . . . cultivated fields have subdued forests: flocks and herds have expelled wild beasts . . . Everywhere are houses, and inhabitants and settled governments, and civilized life. What most frequently meets the view is our teeming population; *our numbers are burdensome to the world* . . . our wants grow more and more keen, and our complaints bitter in all mouths, *while nature fails to afford us her usual sustenance.* In every deed, pestilence, and famine, and wars, and earthquakes have to be regarded for nations, as means for pruning the luxuriance of the human race."[2] (Emphasis added, here and elsewhere.)

These are but a few of the ancients' articulate and impassioned comments and they would probably continue without interruption if we could adequately probe the historical record. Much later, on our own American continent soon after a portion of the human race got a fresh start, the Indian Tecumseh presented "his graceful and majestic form above the heads of hundreds" and, not unlike Isaiah, "made known his mission, in a long speech, full of fire and vengeance. He exhorted them to return to their primitive customs, to throw aside the plough and the loom, and to abandon an agricultural life, which was unbecoming to Indian warriors. He told them that, after the whites had possessed the greater part of their country, turned its beautiful forests into large fields, and *stained their clear rivers with the washings of the soil,* they would then subject them to African servitude. He exhorted them to assimilate in no way with the grasping, unprincipled race."[3]

Probably the native Americans did perceive the problem first. However, shortly after independence from England was declared, Patrick Henry stood before the Virginia Assembly and noted, "Since the achievement of our independence, he is the greatest patriot who stops the most gullies." During the same period, Jefferson worried about soil loss and tried agricultural practices that would minimize such loss. One senses great frustration in the heart of George Washington from a letter written in 1797: "We ruin the lands that are already cleared and either cut down more wood, if we have it or emigrate into the western country. . .A half, a third or even a fourth of what land we mangle, well wrought and properly dressed, would produce more than the whole under our system of management, yet such is the force of habit, that we cannot depart from it."[4]

Nearly fifty years later, in 1845, a Solon Robinson, writing of Mississippi, a thousand miles from Mount Vernon, lamented that "if western land-spoilers knew how eastern land-skinners had skinned their land to death, they would not go on doing just the same thing. But they won't know, and, of course, won't do. . . This is the land of gullies."[5]

The mid-nineteenth century is a great time in American conservation history. Emerson, Thoreau, Marsh and Muir were speaking to the old problems of desecration but with a distinct American flavor. Just as the Hebrew prophets in another land and at another time had given definition to the problem of the loss of a sustainable agriculture and culture, these four men, with knowledge, passion, and style, added their voices and writings at the height of westward expansion.

It was in 1851, before a Concord Lyceum, that Henry David Thoreau delivered his famous, extreme message, "In wildness is the preservation of the world."[6] Like the other three, Thoreau was looking at something broader than the loss of the soils and eventually the loss of agriculture. He was concerned about the very soul of civilization. After a trip to Maine, where wilderness frightened him a bit, he moderated his position on society.[7] Along with Ralph Waldo Emerson he steadfastly contended that wilderness remains the final standard against which to judge our lives.

These two transcendentalists' religious views about wilderness were backed up by a man with a more practical bent, George Perkins Marsh. Marsh's classic *Man and Nature: or*

Physical Geography as Modified by Human Action was published in 1864. Marsh had observed what the "dominion" concept described in Genesis had done and was doing to the American wilderness, and he promoted an alternative approach which sprang from his idea that "Man has too long forgotten that the earth was given to him for usufruct alone, not for consumption, still less for profligate waste." Clear-cutting of the watersheds, Marsh believed, was responsible for the decline of the Mediterranean empires, and, therefore, forest conservation was a practical matter. In this sense he was much like the first head of the U.S. Forest Service, Gifford Pinchot, who believed wilderness should be preserved for economic reasons.[8]

The young John Muir, heavily influenced by the writings of the transcendentalists, hit his own stride in the last third of that century. Being more extreme than either Emerson or Thoreau, to whom he owed a great intellectual debt, he eventually clashed with Pinchot.[9] Muir thought wilderness should be preserved for its own sake and not for economic reasons alone.

Muir, who had gained some fame for his genius as an inventor, eventually became a prosperous farmer.[9] So strong was his love for things wild that he abandoned the life of an inventor, which would have given him both fame and fortune, and also abandoned the life of a farmer, which gave him financial security.[9] In his beloved Sierra Nevada and later Alaska, he feasted his soul on the primitive and pristine. Roderick Nash called Muir a "publicizer" of nature,[10] and one senses that Muir would reach for any straw that might preserve the most wilderness possible. He drew on the ideas of George Perkins Marsh that the soil and forests of the Sierra served as cover in the watershed, which would store water for use below. But this is as close as he ever came to the practical reasons for saving wild things.[11]

There were others around at the same time thinking on the relationship of the human species to nature, but the "critical mass" was built around the insights, eloquence and passion of these four (Emerson, Thoreau, Marsh and Muir). Beyond them, the voice for conservation was but a whisper in those times. And though soil loss was not central to their concern, it was part of the total, for their holistic view gave attention to the connection of everything of earth and its extension into the heavens.

The statements of praise about the wild earth and its limits were by no means limited to mountains or humid regions where the vegetation was lush.

On the Plains, Kansas Senator John J. Ingalls praised grass as the "forgiveness of nature—her constant benediction. Forests decay, harvests perish, flowers vanish, but grass is immortal. Its tenacious fibers hold the earth in place and prevent its soluble components from washing to the wasting sea."[12] Ingalls and scores of others had praised the grasslands with descriptions which made the Great Plains sound like the first Eden. Other area descriptions sounded more like Hell. The fact is that both descriptions are accurate.

The grassland region is given to extremes, and the lack of understanding of these extremes has led to numerous unfortunate problems. At times, stockmen have been justifiably resentful of those who insist that overgrazing has been responsible for all dust storms. Stockmen have overgrazed, but we can't ignore the notes of "the Little Black Father," Jesuit Pierre DeSmet, who described his visit to Montana in 1841–42. "We had to resort to fishing for the support of life and our beasts of burden" he wrote, "were compelled to fast and pine, for scarcely a mouthful of grass could be found."[13] Twenty years later, Jim Bridger noted that the forage was "unusually scarce, and the animals becoming much emaciated."[14] Dust and sand pounded the Conestoga tarps of the new emigrants to Montana in the 1870's.[14] The Dodge City, Kansas, weather station recorded blinding dust storms in 1890, 1892, 1893 and 1894, although overgrazing from cattle had been a severe problem before the collapse of the cattle industry around 1886.[14]

The grass would get dry, but it would come back. The grass would burn and form a black powder for miles, but it would return. In short, the grass withstood all but the plow. Dust storms were not invented with the plow; they simply became more serious because of it. Then there were the years of lush growth due to abundant rainfall, especially in the Great Plains region. This is the complexity the arid lands present us with. Most of the people who came into the area did not understand it. Joseph Kinsey Howard, in his book *Montana—High, Wide and Handsome* tells a story often related in the northern plains country:

> One day in the spring of 1883, as a Scandanavian farmer, John Christiansen, plowed his fields in...North Dakota, he looked up to find that he was being watched...by an old solemn Sioux Indian.
>
> Silently the old Indian watched as the dark soil curled up and the prairie grass was turned under. Christiansen stopped, leaned against the plow handle, pushed his black Stetson back on his head, rolled a cigarette. He watched amusedly as the old Indian knelt, thrust his fingers into the plow furrow, measured its depth, fingered the sod and the buried grass.
>
> Then the old Indian straightened up, looked at the farmer. "Wrong side up," he said and went away.

For a number of years that was regarded as a very amusing story, betraying the ignorance of the poor Indian.

There was a man who understood the region, but he was not listened to either as a prophet or a scientist—John Wesley Powell. Powell became famous for leading the exploration of the Green and Colorado Rivers, but his social and economic recommendations, in many respects, are more important. Powell and his party started with the soil. They sampled the grasses and subjected them to chemical analysis. They recorded the habits of the native plants and carefully measured moisture and penetration in the soil. Various reports of weather observers were synthesized and correlated with the chemical studies of the highly nutritious plants.

Powell left a relatively brief report in the Library of Congress. If it had been heeded as an economic planning document, it would have prevented a great deal of national pain. Powell, the one-armed Civil War veteran, warned the nation that a quarter-section of land alone would be of no value in these arid regions and that the land should not be divided into square miles as it was in the humid East. He suggested sixteen times that amount, four square miles, approximately 2,560 acres, per family. He recommended patterns of settlement organized around available water.[15]

Unfortunately for both the people and the land, the myths were more powerful than the logic. "The rain follows the plow" myth pushed people westward where they would face "unusually dry years," as they saw them, until they would return in their wagons from west of the 100th meridian, offering excuses along the way. Some camped on my grandfather's farm near Topeka and explained that they were "going to Missouri to visit the wife's people for awhile." They couldn't admit they had been dried out and blown away.

Thousands have experienced poverty, hundreds have starved, thousands of livestock have died, millions of acres of soil have been damaged since 1878 when Powell gave his report to a deaf nation. He advised that (1) land surveys should conform to the topography, (2) farm residences would best be grouped (Think of the implications of that advice!), and (3) livestock should run on open range with no fencing. Eventually, settlement and management patterns approached the limits imposed by the region, proving that in the last analysis, land determines. Government policy was slow in catching up to Powell. Sixty years after the report, and nearly four decades after his death in 1902, the government began to catch on. It happened gradually. The Enlarged Homestead Act of 1909, which allowed 320 acres instead of 160 for the plains homestead (still a factor of eight short), required the homesteader to keep one-eighth of the farm *continuously cultivated for crops other than native grasses.*

This was our golden opportunity for sensible land-use planning on approximately a million square miles. As far as land use is concerned, Powell was probably our first scientist and prophet.

In 1906, four years after Powell's death, the book already mentioned in Chapter One, *Man and the Earth* by Nathaniel Southgate Shaler, appeared. Shaler described the importance of a dynamic balance. He cautioned, "The preservation of the food-giving value of the soil as used by civilized man depends on the efficiency of the means by which he keeps the passage of the soil to the sea at a rate no greater than that at which it is restored by the decay in the materials on which it rests." A year later, President Theodore Roosevelt, in a Message to Congress, warned, "To skin and exhaust the land will result in undermining the days of our children."

One hundred and thirty-one years after Patrick Henry declared the gully-stopper the greatest patriot of all, Gifford Pinchot remarked, "The noblest task that confronts us today is to leave this country unspotted in honor, and unexhausted in resources. . .I conceive this task to partake of the highest spirit of patriotism."[5] That was in 1908. Yet, 40 years later, by the mid-part of our century, Aldo Leopold would note that an individual who paid his taxes and otherwise behaved himself was regarded as an upright citizen regardless of what was happening to his land.

Thomas S. Chamberlain, before a 1908 White House Conference of Governors, insisted, "when our soils are gone, we, too, must go unless we find some way to feed on raw rock, or its equivalent. The key lies," he continued, "in due control of the water which falls on each acre. . .The solution. . .essentially solves the whole train of problems running from farm to river and from crop production to navigation."[5]

As early as 1911, Dr. A. M. Ten Eyck of Kansas observed, "The grass roots which formerly held the soil together are decayed and gone, and now, when loosened by the plow, the soil is easily drifted and blown away."[5]

Professor Liberty Hyde Bailey, the grand old man of Cornell University, who lived into his mid-nineties and gave us the impressive *Manual of Cultivated Plants*, sounded a holistic note in *The Holy Earth* published in 1915: "We come out of the earth, and we have a right to the use of its materials; and there is no danger of crass materialism if we recognize the original materials as divine. . .We are not to look for our permanent civilization to rest on any species of robber economy."

Arthur J. Mason was an engineer and pamphlet-writer in Illinois who climbed fences to take samples of prairie soils from railroad rights-of-way and from the adjacent fields. His results on soil loss from plowed fields in Illinois were striking.[16] In 1920 he put into words what later became a much-talked-about consideration on the relationship of land to people. "The instinct that we all feel about good land is sound; perhaps it is a latent feeling that only from good land can a robust stock of men come forth, and one need not go far to verify this. . .I have seen cattle deteriorate in poor country."[5]

One writer, Herbert Quick, had lived during the great transition on the prairie. In *One Man's Life* (1925), he described how he had grown up there on the "old prairie which we feared and loved, and conquered" where there were "beautiful little brook so clear. . .the fisherman's heart seemed to shake the neighborhood. . .We saw no reason for not killing as many prairie chickens as we could, so in winter we trapped them by the thousands. . .I have not heard a bobolink's song for thirty years. It has passed with the little clear brooks and the flights of clamoring wild fowl and all the primitive wildness and beauty and sternness of the prairie."[17]

J. H. Bradley in *Autobiography of Earth* (1935) poetically stated: "The fabric of human life has been woven on earthen looms. It everywhere smells of the clay . . . Howsoever high the spirit of man may soar . . . it is on (the stomach) that humanity . . . ever must advance." A year later, at the height of the Depression, David Cushman Coyle would ask, "Is this the sort of country we want to leave behind us, to be our monument or history? Ravaged land, wasted forests, empty oil wells, and a people starved, stunted, ignorant and savage? Why do we do it? What strange insanity is driving us to destroy ourselves?"[18] In the same year, an engineer, Morris L. Cook, would testify before a Senate Committee that our country is "like a man well gone with cancer or tuberculosis . . . with continuance of the manner in which the soil is now being squandered (we have) less than a hundred years of virile national existence. . .and most difficult of all to change the attitudes of millions of people who hold that ownership of land carries with it the right to mistreat and even destroy their land."[5]

Arthur F. Raper in a *A Preface of Peasantry* (1936), accurately described what we might call the domino effect in southern agriculture. "The collapse of the plantation system," he said, "leaves in its wake depleted soil, shoddy livestock, crude agricultural practices, crippled institutions, a defeated and impoverished people . . . After the boll weevil, the sedge, after the sedge, the silent redeeming pines: and from pines back to cotton . . ." This connection is not so amazing; it only requires some understanding of deteriorating succession and a downward spiral. Raper shows the impact of such agricultural ecology on the human. In a cold and bare rural black school in Georgia, Raper saw these words on the blackboard: "I Opened School, October 15, 1934. Closed School, December 7, 1934. Lord Teach Us To Pray!" The school was open less than sixty days that year: the connection between deprived land and deprived people is immediately clear.

Paul B. Sears, in *This is Our World* (1937), lamented: "Too long we have reckoned our resources in terms of illusion. Money, even gold, is but a metrical device. . .not the substance of wealth. Our capital is the accumulation of material and energy with which we can work. Soil, water, minerals, vegetables and animal life—these are the basis of our existence and the measure of our future."

In the same year, a professor of agricultural chemistry in Wales wrote: "The affairs of the soil may not have the strange magnificence of the outer universe or the curiosity of the inner recesses of the atom; but they touch our daily lives most intimately. So I commend them to your notice and ask your indulgence for the homeliness of my story. . .I propose. . .to tell you something of the tragic happenings to the soil of the United States. They are not without parallel in other parts of the world, but nowhere else has the drama of soil destruction been played so swiftly and on so great a stage."[19]

Others wrote in 1937, at the height of our Depression, frequently calling attention to the parallel between impoverished land and impoverished people. It provided ammunition for Roosevelt liberals who believed that the lack of opportunity breeds sin. In 1938 Sterling North of *The Chicago Daily News* wrote, "Ardently as I have scanned the writings of Europe's half-pint Napoleons I find but one undoubted truth uttered between the two of them. Each has said in effect, 'It takes a rich land to support a democracy.' " North continued, "Every time you see a dust cloud, or a muddy stream, a field scarred by erosion or a channel choked with silt, you are witnessing the passing of American democracy. . .The crop called Man can wither like any other. . . "

In 1939, Secretary of the Interior Harold L. Ickes asked, "Why do we not teach conservation in our schools? Is the waste and pillage and threatened physical destruction of our country less important than the names of state capitals?"[20] The same year President Roosevelt warned, "If we do not allow a democratic government to do the things which need to be done and hand down to our children a deteriorated nation, their legacy will be not a legacy of poverty amidst plenty, but a legacy of poverty amidst poverty."[20] In the same year Roosevelt's Secretary of Agriculture, Henry A. Wallace, told the Land Grant College Association: "Damage to the land is important only because it damages the lives of people and threatens the general welfare. Saving soil and forests and water is not an end in itself;

it is only a means to the end of better living and greater security for men and women."[20]

Hugh H. Bennett, the first director of the Soil Conservation Service, in the first chapter of his *Soil Convervation* (1939), went farther than Wallace by suggesting that land must be loved to be protected. To him the plain truth was that "Americans, as a people, have never learned to love the land and to regard it as an enduring resource." In 1940 E. B. White of *Harper's Magazine* underscored the metaphysical dimension discussed by Bennett. "I can see no reason for a conservation program," he wrote, "if people have lost their knack with the earth." As we have seen in the forty some years since, a conservation program is not enough. White could see "no reason for saving the streams to make the power to run the factories if the resultant industry reduces the status and destroys the heart of the individual." This he called the most "frightful sort of dissipation."

By this time, more and more connections were being made between the health of the countryside and the health of the cities. More began to perceive this unity. Forty years ago, one writer in the *Baltimore Evening Sun* commented that "When a farmer in upland Maryland abuses land, he may be helping to smother oyster bars of the Bay with slime. Many fine oyster beds have been destroyed and the process continues. Land, water and minerals, trees, fish and animals, are all part of the property of Maryland, and it is astonishing how often the protection of one depends upon the protection of some, or all, of the others."[21]

There was a knowledgeable, articulate and impassioned movement under way in the early forties and such men as Lewis Mumford were telling us what must be done to save our country from widespread blight. Our "main handicap will be lack of imagination," said Mumford. "This is one of those times when only the dreamers will turn out to be practical men."[22]

References and Notes

1. Plato, *Critias*, translated by Arnold Toynbee in *Greek Historical Thought*, 1924, pp. 146–147, E.P. Dutton & Co., New York.
2. From the writings of Tertullian as edited by Alexander Roberts and James Donaldson, Edinburgh, 1852.
3. Albert James Pickett, 1971. *History of Alabama*, Vol. II. ARNO.
4. *The Letters of George Washington*, 1966. Doubleday.
5. Reprinted in Vol. 1, No. 1 of *The Land*, Winter 1941, p. 60. I am especially grateful for the chronology provided in this volume, much of which was later printed in *Forever The Land* (reference 12).
6. See Thoreau, "Walking" in Excursions, *The Writings of Henry David Thoreau*, Riverside edition (11 vols. Boston, 1893) or Walter Harding, 1948. "A Check List of Thoreau's Lectures," *Bulletins of the New York Public Library*, 52, 82.
7. "Thoreau's 'Maine Woods' ", 1908. *Atlantic Monthly*, by Fannie Hardy Ekstrom.
8. Pinchot, "Forester and Lumberman in the North Woods" (c. 1894) Gifford Pinchot Papers, Library of Congress, Box 62 as cited in the revised edition of *Wilderness and the American Mind* by Roderick Nash. (1973, Yale University Press).
9. Linnie Marsh Wolfe, 1945. *Son of the Wilderness: The Life of John Muir*, A.A. Knopf, New York, pp. 275–276.
10. See Chapter 8 of the revised edition of *Wilderness and the American Mind* by Roderick Nash, (1973, Yale University Press), for a splendid short chapter on the life of John Muir.
11. John Muir, 1890, "The Treasures of the Yosemite," *Century*, 40, and also "Features of the Proposed Yosemite National Park", 1890, *Century*, 41.
12. Cited in *Forever the Land; A country chronicle and anthology* by Russell and Kate Lord, Harper and Brothers, New York. 1950.
13. H.M. Chittenden and A.T. Richardson, 1905, *Life Letters and Travels of Father Pierre-Jean DeSmet, S.J. 1801–1873*, Francis P. Harper, New York, p. 306.
14. Joseph Kinsey Howard, 1903, *High, Wide and Handsome*. H. Milford & Oxford Univ. Press.
15. J.W. Powell, 1878. "Report on the Lands of the Arid Region of the U.S.", Washington: Government Printing Office.
16. Arthur J. Mason, Autumn, 1950. "Bricks and Hay," Read before the Chicago Literary Club in March, 1917 and reprinted in *The Land* IX:3.
17. Herbert Quick, 1925. *One Man's Life*, Bobbs-Merrill, Indianapolis.
18. David Cushman Coyle, 1936. *Waste*, Bobbs-Merrill.

19. Gilbert Wooding Robinson, 1937. *Mother Earth or Letters on Soil*, Chapter XVI, T. Murby & Co., London.

20. Secretary Ickes asked this question February 27, 1939. President Roosevelt's comment was made May 22, 1939. Henry A. Wallace delivered his remarks to the Land Grant College Association on November 17, 1939. All three comments are quoted in *The Land* Vol. I:1, Winter, 1941.

21. Gerald W. Johnson, Jan. 11, 1940. *Baltimore Evening Sun*.

22. Lewis Mumford, Feb. 1940. *Survey-Graphic*.

4
The Failure of Organizations

The Soil Conservation Service, founded under President Franklin D. Roosevelt was fortunate from the time of its birth. Roosevelt selected Dr. Hugh H. Bennett to be the founding chief of the organization. "Big Hugh," as he was affectionately called, was described by the publishers of the *Farm Journal* as "one of the few immortals of agricultural history."[1] Bennett was not satisfied to simply head another government bureau for he saw his mission to be a "war without guns." The chief encouraged both individuals and groups to support soil conservation work and even went so far as to help establish a citizens society devoted to similar goals for the American soils.

Bennett was one of three people who met on the first of November in 1939 in Columbus, Ohio, to help establish an umbrella organization eventually called Friends of the Land. The other two were Charles Holzer, an M.D. and Brice C. Browning.[2] Holzer had a clinic on the bank of the Ohio River at Gallipolis, Ohio, and had been motivated to do something about soil loss because of the connections he had made in his clinic between washed-out soil and sickly, washed-out people. He donated a thousand dollars to help launch the organization. Browning was active with the Muskingmum Conservancy District.[2]

Bennett was a holistic thinker and doer who felt strongly about saving the back forty. If the back forty were saved, society benefited. To Bennett the entire system—forests and plains, rivers and power, lakes and cities—were one. Bennett had traveled widely and strongly felt the need to make the interreliance of countryside and town clear to all.

People in various government agencies in Washington learned about this meeting from Big Hugh when he returned and they joined him in endorsing the concept, though the bylaws of Friends of the Land prevented their service as officers.

Ultimately, most of the well-known names in conservation were involved in the organization: Liberty Hyde Bailey of Cornell, Gifford Pinchot, the first head of the U.S. Forest Service, J.N. Darling, Paul B. Sears and Aldo Leopold. All wanted the organization to be more than a pressure group. They wanted its members to develop a deep allegiance to sound soil conservation practices.

The organizers circulated a "Manifesto" in 1940 and sought to dispel any idea of threat to the other conservation organizations. Rather, they sought to "support, increase, and . . . unify all efforts for the conservation of soil, rain and all the living products, especially man."[2] The word "ecology" is found but once in the founding manifesto, but most of the founders, whether farmers or professional people, were ecologists at heart. The organization got off to a good start, especially considering these were the early days of World War II. Perhaps the success was partly due to the symbolically significant fact that Dr. Holzer was a physician and Dr. Bennett an engineer. Considerations of both the physical and living world were united as one. The organizational meeting was held in Washington, D.C. with some sixty men and women present.

It must have been an amazing meeting, judging by the group of people present. "There were artists present," recalled Paul B. Sears. A farmer-banker from Missouri had a thirteen word summary of why he was there: "If you get an absorption of water, you prevent an erosion of soil."

The noted author Stuart Chase said at the meeting, "We are creatures of the earth and so are a part of all our prairies, mountains, rivers and clouds. Unless we feel this dependence we may know all the calculus and all the Talmud, but have not learned the first lesson of living on this earth."

This group of people, at once ordinary and extra-ordinary, was advocating a local response to a global problem. That they recognized it as such is clear from the author E.B. White's statement at the meeting: "Before you can become an internationalist you have first to be a naturalist and feel the ground under you making the whole circle."[2]

At the third annual meeting, in 1942, the international theme surfaced again. Hugh Bennett was just back from visiting soil conservation-trained people in Ecuador and Mexico. He commented that "when you get out on the land with people, and work with them and talk with them about the productivity of the soil, there is some sort of common denominator there. I think that our statesmen, our educators and all of our great men from the beginning of time, have missed that point.

"When you begin to talk and work with the fertility of the soil and the way it relates to the welfare of humanity, you are talking a common language. It brings people closer together. It will bring nations closer together."[2]

At the same meeting the international need was further articulated by Dr. John Detwiler, President of the Canadian Conservation Society. "We begin to realize," he began, "that an overcrowding of people on a diminished soil base may impinge on the intellect, lead to physical and nervous disorders, and break forth ultimately in the hidden hunger that brings on wars. Perhaps when we organize conservation on an international basis we can avoid the hidden hunger which brings on wars." In 1945, Detwiler sent a communication to the organ of the organization, *The Land Quarterly*, which included, "To preach conservation at such a time, when all our resources, national and otherwise, are being sacrificed in unprecedented measure, might seem to some anomalous, even ironical . . . But we firmly believe, and now are more acutely aware than ever that conservation is basically related to the peace of the world and the future of the race."[2] The journal carried an "Other Lands" section as further indication of the group's awareness of the connection between prosperous soil and peaceful people.

From the outset, the leaders, many of them government workers, recognized that *The Land* "should be written from the ground up and not from Washington down." Symbolically they moved the central office from Washington to Ohio in 1941. It was about this time that Louis Bromfield of Malibar Farm joined the organization and devoted much time, money and his own literary skills to the organization and the cause.

Let us consider for a moment the organization in *government* these people were trying to support. Bennett had quickly assembled the most able engineers, agronomists, nurserymen, biologists, foresters, soil surveyors, economists, accountants, clerks, stenographers and technicians of many backgrounds into the service of saving the soil. There was an anxious urgency on the part of nearly everyone involved. The Soil Conservation Service was, as Wellington Brink put it, "born with pride and loyalty and a sense of high destiny—an inner element which was to persist and spread and animate the organization and weld it together with a spirit altogether unique in modern government."[3] Because of the high caliber of the people employed, the SCS gained a good reputation fast.

The SCS was an organization based on four steps: science, farmer participation, publicity, and congressional relations. Bennett explained these four steps in his characteristic North Carolina accent many times. "By science," he would say, "we tried to imitate nature as much as we could. We abided by the following basic physical facts, (1) land varies greatly from place to place, due to differences in soil, slope, climate and vegetative adaptability; (2) land must be treated according to its natural capability and its condition as the result of the way man has used it; (3) slope, soil, and climate largely determine what is suitable protection in all situations . . . Above all . . . we tried to imitate nature."[1]

The farmer participation step was extremely successful in helping farmers become interested and involved. SCS men would walk over a farm, field by field, with the owner. There were tens of thousands who came to the famous soil conservation demonstration meetings. The entire approach was touted as a most democratic and practical approach to encouraging farmers to safeguard the natural resource. When Bennett retired, over one and three-tenths billion acres were in the program, and it involved nearly five million farms. And here is a point we need to come back to later on. The per acre crop yields began to rise steeply between 1935 and 1950. Because there was a correlation between sharp crop production increases and the increasingly widespread soil conservation practices, it appeared that soil conservation was the cause. But, again, more on this later.

Bennett's third approach, publicity, seems to have been fully exploited. In his own words: "We employed capable writers, trained journalists who knew what to write and where to get articles published. They were not agronomists or agricultural engineers trying to be reporters or advertising men. They were journalists who knew good pictures, good news angles, good feature possibilities when they saw them. And I let them go at it, in their way, which happened to be the newspaper and radio way. The country papers picked up our . . . material by the volume and it got back to congressmen . . . Clippings piled up, requests for more information poured in . . . We considered the press and radio a very important part of our job." USDA records will show that Bennett was right.[1]

In the last step, congressional relations, Bennett's charming and colorful personality ensured success. He kept the congressmen informed of the conservation effort in their home districts and talked to them about plans for the future. Bennett became a legend on the Hill. "When I appeared before a committee," he explained, "I never talked about correlations or replicas. But I did spread out a thick bath towel one day on a table before a committee, tipped the table back a bit, and poured a half pitcher of water on that towel. The towel absorbed most of the water, cutting its flow from the table to the rug.

"I then lifted the towel and poured the rest of the pitcher on the smooth table top, watching it wash over the edge onto the rug. I didn't say anything right away—just stood there looking at the mess on the floor. Then I looked up at the committee and explained the towel represented well-covered, well-managed land that could absorb heavy, washing rains. And that the smooth table top represented bare eroded land, with poor cover and management on it. They seemed to understand, because we got our appropriation."[1]

This was the organization and the chief that the Friends of the Land sought to support.[4] The society was itself free of government, unpretentious, informal, friendly, devoted to "rational evangelism" with high-level discussions and lots of adult education. Their mission was education and they believed in the democratic way, sometimes (one is tempted to think) to a fault. One member once wrote that "no lasting reform can possibly be accomplished by compulsion, by bribery, or by regimentation. Information, knowledge, *education* that extends to all callings and all age groups of the people—that is the only democratic way."[2]

In a review of their works and aims, their editor commented, "It was not our purpose to conduct lobbies, either in Washington or at the state capitols, but rather to attack by meetings, tours, institutes, publications, press dispatches, articles, books, movies, radio and all educational media a prevailing public ignorance and inertia."[2]

"This is slow but it has the advantage of being a frontal assault," a friend wrote in the *Baltimore Sun*. "If carried through, [it] will be final. For if the American people were actually informed and alert on conservation, then programs of legislation would be merely matters of detail."[2]

I suspect that never has such a large group of people seen so many links in the cause-effect chains of nature. On the cover of several issues of *The Land Quarterly*, a well-written and artistic publication, is the statement, "A Society for the Conservation of Soil, Rain and Man." Aldo Leopold, who wrote articles for the journal, reminded readers in his rich prose of the connections between people and land. It was he who wrote in *A Sand County Almanac*: "Health is the capacity of the land for self-renewal. Conservation is our effort to understand and preserve this capacity."

The organization was phased out in the late 1950's, but their publication revealed that it was sinking before then. How could such an organization go under, an organization which held hundreds of meetings, often at summer camps, at teachers' institutes and at metropolitan centers? They traveled to the hinterland and held public forums. They were not cultists or fanatics. They simply wanted to explore the facts and pass on the information. It was typically American, like the Farm Bureau or the Chamber of Commerce. It was not nearly as sexist as many organizations were at that time. In 1949, the fifth president, upon taking office, said, "The care of the earth for all the years to come is not exclusively a stag affair."[2]

The organization probably went under because of the nature of the 1950's. It was an era of great complacency in America. Technology was solving our problems. Everything was going to be bigger and better and the opportunities were limitless. It was not a time when many were inclined to listen to the dire warnings of a few who happened to be reading different signs of the time.

Standing Back for a Moment. If the year were 1930 A.D., one might conclude that Job and Isaiah and all the other Hebrew prophets were early voices of a minority that had extremely sharp vision. Patrick Henry and George Washington and Thoreau and Emerson, Muir and Marsh were only the perceptive and eloquent observers on a continent that was then regarded as yet young and endless in its resources. One might contend that all these greats constituted the important historical stream of pioneers upon whose shoulders a movement finally stands.

By 1940 A.D., with the Soil Conservation Service and supporting organizations, we might be comfortable that it is all finally coming together. Government, industry, and private organizations are winning the war against soil loss.

By 1950 A.D. we who are given to examining correlations nod our heads knowingly and restfully as we note that gullies *are* arrested and *production has increased twenty, thirty, forty, sixty and even one hundred percent.* Other governments would now follow our model.

We now come back to the discussion tabled earlier in this chapter pertaining to the dramatic yield increase between 1935 and 1950. Much of the per-acre increase was due to retiring marginal land from production. For example, if we have two acres of corn, one which produces 120 bushels per acre and another which produces sixty, the average would be ninety bushels per acre. By retiring the low-producing acre, we immediately experience a thirty-three percent increase in per-acre production.

For purposes involving federal budget allocations alone, one would expect SCS officials to take as much credit as possible for the increases in crop production even though the farmers who were paying for the fertilizers and pesticides might have thought otherwise. To the extent SCS scientists were taking credit for production increases, it is to that same extent that they were ignoring one of the oldest warnings of science: scientists should be careful in their talk about cause and effect and speak first and foremost of correlations. For, if we are to take seriously the impressive studies from the early 70's to 1976, annual soil loss is greater now by at least twenty-five percent than in the Dust Bowl years when the SCS was begun.

It turns out that any correlation between higher crop production and the activities of farmers through the Soil Conservation Service is like the often-cited positive correlation between the number of bottles of whiskey and the number of Bibles sold in Canada. The SCS certainly did help the country in setting aside marginal land. Furthermore, I would hate to think what would have happened to our soils without the SCS.

We cannot assume, however, that SCS-supported activity is the primary cause for higher production, except in a few places. We should look to the increase in fertilizer, pesticides, improved varieties and irrigation. This is a very important point, for as I have already stressed, our fossil fuel-based chemotherapy prevented us from seeing the wasting away of our soils and gave us an unrealistic view of actual soil health.

The painful conclusion we must live with is that both governmental and citizens organizations have failed to prevent soil loss.

Can we expect the 1977 Soil and Water Resources Conservation Act (RCA) to succeed where past action has failed? This Act is cited by the USDA as the most important legislation since the first soil and water conservation measures were implemented in the 1930s.

The USDA is proposing a major overhaul of programs throughout its 34 separate administrative offices in order to meet anticipated demands on the nation's natural resources over the next 50 years.

The implementation of the legislation began with public hearings in 1978 to identify problems in the conservation effort. I went to the one held in our district. It was poorly attended; according to one official at the meeting, only one or two people represented some of the districts. The meeting was very disappointing and I got the feeling one experiences at such meetings, that it was being conducted more out of simple compliance than out of any earnest effort to elicit spontaneous elaboration about what was wrong with the conservation program. I suspect that most meetings, all across the country, went this way. Nevertheless, according to the summary statement of all these meetings, the most often-repeated complaint was that farmers needed more technical help from the Soil Conservation Service in order to both plan and implement conservation programs.

Following the nation-wide public hearings in 1978, seven resource areas were identified and objectives for each of the seven were established. Next the USDA proposed seven conservation strategies to meet the proposed objectives. At the hearings farmers, ranchers, and interested citizens were asked to consider the proposed policies, which range from totally voluntary compliance to complete regulation. One extreme regulatory measure, if it were approved by the President and Congress, would require landowners to apply acceptable conservation measures or be denied any assistance from USDA farm programs.

The potential effectiveness of any new program in the conservation of soil and water will depend not only on the new rules handed down by government, but on the *owners* of the land. New ownership patterns have been emerging rapidly over the last 20 years. Over half of all agricultural land has changed owners since 1960 and now 2–3 percent of all agricultural land changes ownership each year.[5] There have been drastic changes in land tenure. Land has become increasingly an item for speculation and absentee ownership is widespread.

There is another reason the RCA effort will be hampered, perhaps the most important one of all. There are fewer families on the land and consequently fewer people who will love the land in any way close to the way it deserves and needs to be loved. RCA administrators have already anticipated this reality and expressed it in their 1980 review draft.

My own hope is that the RCA program will raise the level of consciousness about the need to conserve. But my bet is that in 10 years soil loss will be greater and the amount of energy going into agriculture will have increased. Increasingly, prime farmland is being taken out of production by urbanization and industrialization. This forces a shift to more intensive agriculture on non-prime acres. This results in reduced production, even when more energy-intensive fertilizers are used. It also means increased soil loss, for non-prime land is usually steeper. In Black Hawk County, Iowa, for example, prime farmland averaged 106 bushels of corn per acre in 1978, 5 bushels more than the national average that year. On non-prime soil, the average yield was 66 bushels! Soybeans in that county yielded 40 bushels on prime land but only 25 bushels per acre on non-prime land.[5]

The US, relative to its level of consumption, is resource poor outside of forests and soils. The temptation for any Administration will be to spend these resources to minimize the balance of trade deficit. In 1977 the cropland base was 413 million acres. According to the National Resource Inventories nearly one-fourth of this acreage, 97 million acres, has eroded in excess of 5 tons per acre per year on average.

We can applaud the efforts of the RCA administrators, but the longer perspective teaches us to not expect much. Stewardship has simply been inadequate, as we saw in the Prologue.

Short run success in production has contributed to our arrogance and has blinded us to the fact that an engineering approach to what is basically a biological problem won't do, as we saw in Chapter 2. The modern soil scientist has been as ignored as were the prophets of the past, as we saw in Chapter 3. Finally, the collective efforts of the largest government-supported soil conservation organization in the history of the world, the SCS and the supporting citizens' organization, Friends of the Land, as hard-working and dedicated as they have been, have been inadequate to the task.

The problem of agriculture remains and, as I will contend in the next chapter, it could get much worse.

References and Notes

1. Sanford Martin, Summer, 1959. "Some Impressions of Hugh H. Bennett, Father of Soil Conservation." This biographical sketch was first published in *Better Crops with Plant Food* by the American Potash Institute. It was later reprinted in *Land and Water*.

2. Russell Lord, Winter, 1953. "The Whole Approach," *The Land and Land News*, Vol. IX:4. This issue was a special review and preview for 1941–1953.

3. Wellington Brink, 1951. "Big Hugh's New Science." *The Land*, Vol. X, no. 3.

4. For a more extensive coverage of The Friends of The Land, see Russell and Kate Lord's book *Forever The Land; A country chronicle and anthology*, Harper & Brothers, New York. 1950.

5. Summary of Appraisal, Parts I and II and Program Report, Soil and Water Resources Conservation Act (RCA), Review Draft, 1980. USDA. Also Review Draft Part I.

5
Food as Fuel: A Prophesy of Collapse[1]

With a little planning it will be easy to meet our basic shelter, food, and clothing needs in the 21st century with sunshine. For example, many new solar homes designed from the ground up have fewer leaks, better insulation and more square footage than the owner had in his or her previous home. In this shift, quality of life should not decline—it may even be enhanced—and it will simply be a matter of time before solar space and hot water heating are available to everyone. Life as usual. Fifty years from now we will realize it wasn't all that painful. From the perspective of one generation, the shift was really quick and easy. So much for the problem of providing warm shelter with sunlight. Next problem?

Food supply in the coming new age of limits need be no problem at all, once we realize that the energy requirements of food production at the field level are very small. Future government rationing plans need only guarantee a plentiful supply of liquid fuels to the farmers and they can continue to guarantee an abundance of food, using 200 horsepower tractors complete with air conditioning and stereo. Why not? The distribution of this food may be another problem, but more about that later.

The third old-fashioned basic need—clothing—will be met as it has been in the last 50 years. Even if we continue to make clothing out of oil, it is better to do that than burn that oil—even in an economy car. We need not be naked and cold during the coming age of energy scarcity.

The industrial revolution has greatly expanded a fourth type of human activity and turned it into a need—transportation. We used to walk (powered by solar energy stored in our food) or ride in a buggy, wagon, or ox cart (again powered by solar energy stored in the food the animal ate). In short, if we had food and were protected enough (with clothing and shelter) to remain healthy, we could transport ourselves. Ordinary walking was part of living, but a journey was an event that must be talked-up and planned. By contrast, the 15-mile commute to a job is like a 15-minute walk to the fields outside the European village. To deny individuals the opportunity to be transported to their work is to deny them of their right to earn a livelihood. It is a necessity.

Energy From the Fields. It is the expansion of transportation to a necessity ranking with the old basic needs of food, clothing, and shelter that has created a profound problem for the landscape. Here the limits of a sunshine future become most clear. I am *not* talking about how terrible it is that roads gobble a million acres of land each year or lamenting that the average interchange uses 40 acres. Instead there is a need to be concerned with the difficulty of making the shift from portable *fossil* fuels to a sustainable source of portable *sunshine* fuels derived from biomass.

Around twenty-six percent of the total energy demands for Americans is in the transportation sector.[2] Despite the recent euphoria in the Department of Energy over electric cars, we can discount their importance, for they must be charged or run on electricity probably generated by a nuclear or coal-fired generating plant. The society is becoming rapidly aware of where the ambiguities lie in nuclear power, and much too late for grown people have realized that before nuclear power is feasible, everything has to work. We are finally admitting as a culture that such a guarantee cannot be made to the satisfaction of enough people. When the costs for reclaiming strip mine areas for western coal begin to come in we may discover it is not the bargain it appears to be now. Since battery storage is materials-intensive and clumsy, and since strings of electric lines on the median strip of the interstates as well as along side streets present problems hardly worth fooling with, we are forced to turn back to liquid fuels, which have traditionally met our transportation demands. Only a few besides the politician who thinks ahead only to the next election, or those who stand to profit financially, can be very happy about our currently proposed crash synfuel program, which seems bent on taking a 200-year supply of coal and converting it to an ephemeral 40-year supply. When the acid rains drench the soybean fields, even agribusinesses with oil interests may begin to develop second thoughts.

Methane gas and methanol derived from manure and other waste products offer possibilities, but are in extremely limited supply, given our level of energy consumption. Some are insisting that alcohol, whether it be derived from wood products, garbage, grains or sugar beets, is our best bet at the moment.

Indeed, a future society based on sustainable energy requires that we look to a sunshine future and that we regard all fossil fuels as transition fuels. Such an energy future will likely use an approach in which numerous energy sources are scaled to meet numerous end-use needs. We will have to use these sources of energy at the regional level to meet our needs: direct solar photovoltaic, wind, microhydro (including low-head run-of-the-stream), and biomass conversion of plant materials and wastes.

We should acknowledge from the outset that extensive use of any one of these partial solutions will not accomplish what a *moderate conversation* program will.

The Land's Potential for Liquid Fuels

What kind of potential for fuel production does the American land hold? Plants collect sunlight at very low efficiencies and store it in a high quality, energy-expensive chemical form. We can easily calculate the land's potential for production of portable liquid fuels based on our country's available land area. If every acre of the coterminous 48 states were suddenly to become as productive as our average corn field was in 1978, a little over 100 bushels per acre, the alcohol yield would be about equal to our current transportation fuel use. And we'd have to grow these corn plants on every acre of deserts, mountain tops and cities.[2]

Since this is not possible, what is?

In terms of total crop production, corn is our top potential alcohol producer. Sugar beets could out-do it and plants such as sugarcane have an even higher alcohol yield, although only by a factor of two or three under favorable tropical conditions.

If we do an energy balance analysis to see how much energy goes into alcohol production and how much comes out, the results are disappointing, even when we are overly generous.

The most persuasive proponents of alcohol fuels usually assume that the grain carbohydrate will be the source of alcohol. They propose that the left over protein mash be mixed with stalks (or stover, as it is called) and fed to livestock without affecting the animals' weight gain. In the case of the corn plant, half the total biomass is in the stalk.[3] To feed the stover along with the mash is naturally appealing; on the surface, we are home free. Not only do we have fuel to run our transportation system, we have recovered a former waste product and turned it into a useful resource, cattle feed.

The mistake is in regarding the stalk as waste. It isn't good enough to simply return the nutrients in the stalk to the field. Organic matter is important in the soil. Nutrients in organic matter are tied up in "time-release capsules," set to coincide with the rhythms of the growing season. Nutrients from commercial fertilizer, by contrast, have to be supplied in greater quantity and at a larger expense in fossil-fuel derived energy. In addition, organic soils have more porosity and therefore better water-holding capacity and gas exchange. This is a way of saying that these soils are in better balance and health. Part of the problem is that we don't see the organic carbon in the soil and tend not to appreciate what is involved in its formation or how long it has taken to accumulate. Though it varies from soil to soil, the average age of soil carbon is in the hundreds of years.[4,5] Some organic molecules have been discovered to be 3,000 years old![5] And they are still at work.

Assume 150 tons of soil per inch per acre with the total organic matter in the first foot or so amounting to 6–8 percent, on the average.[5] We can safely assume 9–10 tons of organic matter in the first inch. Even though organic matter declines steeply as we go down, clearly there is more organic matter below the soil line than above it.[6] An alfalfa field will produce about seven tons a year and a corn field around 10 tons.

What does this have to do with an alcohol fuels program? The plowing of the grasslands has already cost the prairie soils about 30 percent of their organic matter.[5,7] (It would be interesting to know how much of the carbon now present in the atmosphere is the consequence of opening up North America.) Amory Lovins has called this organic matter "young, dispersed coal."[8] It continues to disappear for, even though massive doses of inorganic nitrogen have been applied in the Corn Belt, soil nitrogen and organic material are declining at 0.5 percent a year.[9] This may reach equilibrium eventually, but how much organic matter will be left and what will productivity be like when it does?

A massive alcohol fuels program could turn out to be, however unwittingly, a harvest of young soil coal. Unlike real coal, these are working molecules and their energy value in the soil is far greater than their equivalent in a fuel tank. If organic matter is continuously mined from the field, ecological debts will accumulate to the point where the tenant might be ousted from the land without the dignity of an eviction proceding. Stover is not waste, and, over the long run, must be left in the field.

So where does this leave us? If we imagine a massive alcohol fuels program we must consider humans and their livestock to be in competition with the automobile for *grain carbohydrate* that exists now—or more land must be brought into production. What is the land's potential now for liquid fuels? For our purpose let us assume no energy input into the alcohol plant; it is to be built from local materials and operate with a previously unused energy source, such as direct solar, wood or wind-electric. Let us further assume that the spent mash (distiller's grains), which is a protein by-product, can be substituted for other protein sources. The mash therefore is given an energy credit equivalent to the quantity of energy that would be required to grow the same amount of protein elsewhere. Using the 1978 U.S. average yield of 100 bushels per acre for corn, we might expect to harvest 84.6 gallons of ethanol per acre with on-farm alcohol production, after we subtract the energy costs of the fertilizer, pesticides, machinery manufacture, traction for seed bed preparation, maintenance and harvest, and add the spent mash energy credit.[2,10] Most (97.6%) of this total equivalent energy yield comes from the credit we give the spent mash by-product (i.e., the energy freed from soybean production). This is a disappointing result, for this 84.6 gallons per acre yield has an energy value equivalent to only 1.2 barrels of crude oil.[10]

Less energy is needed to grow wheat than corn and the higher protein content yields a larger credit. Nevertheless, assuming the 1977 wheat yield of 30.6 bushels per acre, there would be an optimistic alcohol harvest of 70 gallons per acre—the energy content of which is equivalent to 1.02 barrels of crude per acre.

These are sobering results, for if enough agricultural alcohol is to be produced to make a dent in U.S. oil consumption, millions of acres of currently unfarmed land must be brought into production. Farmers are already cultivating their most productive lands[11] and any new lands will be less fertile and of a steeper slope. *These are lands in which the erosion will be greater than the national average, lands in which fertilizer inputs will be necessarily greater than the current national average—and the yield will often be less than the current average.* As mentioned in the last chapter, in Black Hawk County, Iowa, prime land produced 106 bushels per acre of corn in 1978 while marginal land managed only 66 bushels per acre on the average. In the same county, soybean yield was 40 bushels per acre on the good land, and 25 on the marginal land.[12] Consequently, the average energy balance will be even less on a per acre basis. Remember that the General Accounting Office's study of 1976 found the average soil loss in the Corn Belt, Great Plains, and Northwest to be around 16 tons per acre per year.[13] An Iowa State study in 1972 set the figure at 12 tons per acre per year for the nationwide average. Let us be conservative for purposes of illustration and assume 9 tons of soil loss from a typical field used to grow 100 bushel-per-acre corn to make alcohol.

Let us pose a very simple question. At 9 tons of lost soil per acre, how many pounds of soil does each gallon of alcohol represent that we send from our corn fields to the transportation sector? It can be determined as follows:

$$\frac{\text{lbs of soil loss}}{\text{gallon of alcohol}} = \frac{9 \text{ tons}}{\text{acre}} \times \frac{2000 \text{ lbs}}{\text{ton}} \times \frac{1 \text{ acre}}{100 \text{ bu}} \times \frac{1 \text{ bu}}{.846 \text{ gal}} = \frac{213 \text{ lbs}}{\text{gal ethanol}}$$

With the addition of marginal cropland to the productive sector, we can estimate that the average loss is increased to 9.26 tons per acre.[14] This may not seem like much, only six more pounds of soil per gallon for a total of 219 pounds per gallon. It is important in the analysis of alcohol production to look beyond the individual consumer and producer to the impact on the nation as a whole.

We currently farm 400 million acres in this country, and since we have an average loss of at least 9 tons per acre, the total amounts to 3.6 billion tons of soil loss per year in the U.S. (This is certainly conservative since the Iowa State University study concluded there were 4 billion tons of soil lost each year.) Let us look at a comparison:

Current lands: 400 million acres × 9 tons/acre = 3.6 billion tons
Current land + marginal lands: 511 million acres × 9.26 tons/acre = 4.7 billion tons.[15]

This amounts to a 30.6% increase in erosion nationwide! Here we are assuming the current crop mix. With a massive alcohol program, there will be even less incentive for crop rotation, even though in Missouri continuous corn land will lose as much as 20 tons per acre, while similar land in rotation with wheat and clover will lose only 2.7 tons per acre. If alcohol fuel crops do what the farmers want and need—raise crop prices—then all of the usual cost/benefit analyses will suggest that the farmer should "pull out all the stops" for alcohol production. As with nuclear power, the subsidies will hide the true cost and totally obscure the fact that our precious dollars are promoting thermodynamic tail-chasing. Soil will disappear from the American fields in the same era our coal is being wasted in the synfuels program. If the Carter administration, and others that follow, promote action to match their rhetoric, and actually develop massive alcohol fuels and synfuels programs, both soil and coal will disappear at about the same time—around 50 years from now.

We should not be deeply concerned that the nation had to spend tens of millions in 1975 to dredge and remove sedimentation from the nation's rivers, lakes, reservoirs and ditches. Nor is it very important that the same job today would cost at least a third more, since we must consider the marginal land brought into production (30.6% more erosion). This sediment has lost most of its value as productive topsoil, so the dredging merely represents another external cost to society from erosion.

However, there is a more fundamental problem. The USDA has estimated the annual cost of replacing the nitrogen, phosphorous, and one fourth of the potassium lost through soil erosion at $18 billion per year in 1979 dollars.[16] If we bring marginal land into production, replacement fertilizer will cost 30% more. Again, the dollar cost is not of particular importance.

The fact of most importance is that fertilizers are fossil-fuel-based and that we have literally moved our agricultural base from soil to oil.

It may be appropriate for us to promote some alcohol production for farm use as part of our multiple, or "fine-grained," approach to a sustainable energy future. Even so, our fields will be pressed if we try to meet only our direct farm energy needs. Gasoline and diesel fuel for farm production account for approximately 6% of the fuel consumed by all motor vehicles. To meet this need alone, we would need approximately 140 million barrels of oil equivalent.[17] This could be provided by 117 million acres of corn for alcohol. In 1977, *70 million* acres of corn were harvested!

We should keep in mind that the energy in the alcohol required to meet the demands of an average U.S. car for one year could alternatively be used as food to feed 23½ people for an entire year.[18] The issue is not whether the alcohol is there, but that massive alcohol production from our farms is an immoral use of our soils, since it promotes their rapid wasting away. In the longer run—say 20 years—it will be easier to deal with the political realities associated with scarce energy for a wasteful transportation system than the political realities associated with a scarce food supply and extremely high prices for what food is available.

Much of the current clamor for alcohol fuels is to provide the fuel farmers require for production (gasoline and diesel only). To meet this requirement, we would have to plant corn over 117 million acres (the 70 million acres now in corn plus 47 million acres of marginal land) to make enough alcohol. Because we have credited the spent mash, none of this would go into livestock or people. But as bad as the results of these optimistic assumptions are, it becomes even more bleak as we get closer to reality. To be closer to reality we will probably have to assume that we must plant the current 70 million acres to corn *plus* the 111 million acres of marginal land, because this latter is of such poor quality and steep slope. If we take this route, we can meet the on-farm needs and increase the erosion by 30.6%. On the other hand, we can increase the average number of people per vehicle per mile from 2.2 to 2.4, thereby releasing 140 million barrels of oil from automobile consumption. We must devote more effort to looking for the most appropriate mix of energy sources, including conservation.

With appropriate planning we can expect the energy transition to be rather untraumatic in providing the age-old basics of food, clothing and shelter, including staying warm. But in the area of food for transportation, we can summarize with what, at first glance, may appear to be three seemingly wild assertions.

First, our high-yielding cereals, as products of technology, have already destroyed more options for future generations of Americans and, indeed, people all over the world, than the automobile.

Second, if cereals are converted into ethanol or grain alcohol to run cars as fuel-hungry as ours are today, on any meaningful scale, we will have committed ourselves—unwittingly or otherwise—to a serious moral decision. The churches have not even begun to perceive, let alone address, this problem. Yet, historically, with our abundance of food, we have been generous. We have worried about the starving Armenians, Cambodians, Biafrans, and Boat People. Historically we have asked, "How can we make our plenty available to the people who need it?" Of course the cynic can offer that our concern is basically economic and that any morality is but a thin, "stained glass," veneer over a crass economic interest. The cynic may be right and if he is we will soon know. If we embark upon a massive ethyl alcohol production campaign to feed America's automobiles from food, we will quickly see how thin that veneer is. We will no longer need to worry about the problems of making food available to those who need it for food.

The third assertion that naturally follows is that if we do launch a massive campaign to meet even a small portion of the current demands of the automobile, some of us and most of our children will forever pay for that sin in options lost with the eroded soil.

These are strong assertions but it may not be difficult to prove they are not strong enough.

Third world people will not be able to outbid Americans who want that food for their cars. There will always be a plentiful supply of people who feel important enough to travel over 500 miles an hour. Many are in positions associated with power.

When the decisions are made as to whether so much acreage is allocated to the growing numbers of poor people for food or fuel for their own movement over the surface of the earth, there is likely to be a large number who will feel that their trips are "necessary for the economy."

The soul of a people may best be described by how much of the future they are willing to wittingly discount. Numerous upright citizens, with the best of intentions and with some degree of justification, can accept the erosion of sustainability in the materials sector of our economy because of their belief in the cleverness of our technological community to find acceptable substitutes. "When one metal is gone, a synthetic product will take its place." One can have little enthusiasm for such a faith in the technological fix, but if those optimists are wrong, the consequences may or may not seriously affect the food system. But if the optimists are wrong about the capacity of the land to fuel a terribly wasteful transportation system, then agricultural disaster is imminent.

We must preserve our soils for continued use as a food-giving resource, to produce products that formerly elicited prayers of thanksgiving at meal time, realizing that the soil and not oil or its substitutes is the basis for the long-term health of the civilization.

Certainly we should expect some alcohol production from garbage and other biomass, including crops, but a *massive* program from our fields must be stopped. So what is the answer for transportation fuels? Environmentalists have gone along, correctly lamenting the wastefulness of our society, noting the gains that will come with more efficient ways of doing our tasks, comfortably acknowledging that they must be done. We have said to ourselves that driving 10,000 miles every year is all right, if only we can get 75-mile-per-gallon vehicles instead of roaring around at 17 mpg. But a more realistic assessment suggests that we must strictly increase efficiency in our transportation machines, we must car pool and we must radically cut our total travel. These last two involve dramatic changes in lifestyle and if we don't impose them upon ourselves, if we continue as we are now, our future will include electric lines running down the median strips of our highways, leading back to nuclear and coal-fired power plants, and the sunshine future will be impossible.

Over 100 years ago, John Wesley Powell told America to expect little of the lands west of the 100th meridian. But our nation was young and deaf to anyone who believed that, in the long run, *land determines*. Americans believed at our first centennial, as they do now, that ultimately *people determine*. That deafness has proved destructive to both people and land. But the ecological and human consequences could be even more disastrous today: there are more people involved and a larger percentage of our highly productive agricultural landscape will be damaged. Powell is still right.

References and Notes

1. Much of this chapter was presented before the National Alcohol Fuels Commission in November, 1979. A similar version was presented to a DOE gasoline rationing panel in Chicago in early January, 1980. Both papers were a cooperative effort among Mari Sorenson Peterson, Research Associate at The Land Institute, Salina, KS, Professor Charles A. Washburn, California State Univ., Sacramento, and myself. I thank both of them for their substantial contribution to this chapter.
2. Charles A. Washburn, Fall, 1979. "Energy from the Land," *The Land Report*.
3. "Sandhill Farmers succeed with minimum tillage," May-June, 1980. *Irrigation Age* pp 12.
4. E.A. Paul & W.B. McGill, 1977. "Turnover of Microbial biomass, plant residues & soil humic consituents under field conditions," in *Soil Organic Matter Studies*, Vol.1:149–157.
5. George W. Cox & Michael D. Atkins, 1979. Agricultural Ecology: An analysis of world food production systems. W.H. Freeman, San Francisco pp 271–272.
6. M. Schnitzer, 1978. "Humic sustances: chemistry and reactions," in *Soil Organic Matter*, M. Schnitzer & S.U. Khan (editors) Elsevier Scientific Pub. Co., NY.
7. C.A. Campbell, 1978. "Soil organic carbon, nitrogen & fertility," in *Soil Organic Matter*, M. Schnitzer & S.U. Khan (editors) Elsevier Scientific Pub. Co., NY.
8. Personal communication.
9. B. Commoner, *A Study of Certain Ecological, Public Health and Economic Consequences of the Use of Inorganic Nitrogenous Fertilizers*, First year progress report to the office of Interdisciplinary

research, NSF, submitted by the Center for the Biology of Nat. Systems, Washington Univ, St. Louis, Mo, 5 May 1972.

10. One bushel of corn will yield 2.2 gallons of alcohol. An acre producing 100 bushels will thus yield 220 gallons. Using calculations based on Pimentel and Terhune ("Energy and Food," Annual Review of Energy, Vol. 2, 1977) we must subtract 218 gallons per acre in alcohol equivalent as the energy cost for producing the crop. Thus, 220 gallons −218 gallons leaves a net of 2 gallons of alcohol equivalent. But when we credit the protein mash with the direct and indirect energy it would require to grow a substitute protein equivalent in the high-protein soybean crop, we have a net of 84.6 gallons per acre.
11. "Corn Yields . . . Where Do We Go From Here?," by John Marten. *Farm Journal*, May 1979.
12. Soil and Water Resources Conservation Act, Summary Appraisal, 1980. Review Draft, Part I and II Page 5.
13. "To Protect Tomorrow's Food Supply, Soil Conservation Needs *Priority* Attention," General Accounting Office, March 1, 1977.
14. Marginal land's potential for conversion to cropland is categorized as high, medium, low, or zero by the Soil Conservation Service (*Potential Cropland Study*, October 1977). Marginal land is defined here as that in the high and medium categories, since only this land can reasonably be brought into production. This amounts to 111 million acres in the United States. The SCS categories are based on the economic feasibility of bringing the land into crop production considering development costs, production costs, and commodity prices in 1974.
15. This analysis is based on data in the *Potential Cropland Study* by the Soil Conservation Service. Of the marginal cropland, 32% of the acres have severe erosion and 27% have above average erosion problems. We can conservatively assume that 12 tons of soil loss per acre represents severe erosion and 10 tons per acre represents above average erosion. Using a weighted average formula, agricultural land experiences 10.2 tons per acre of soil loss on marginal land.

    With all high and medium potential cropland brought into production, we would be using a total of 511 million acres. Averaging current cropland (400 million acres) at a soil loss of 9 tons per acre and marginal cropland (111 million acres) at 10.2 tons per acre, we arrive at an average of 9.26 tons per acre of soil loss with all current and potential cropland in production.
16. President Carter's Second Message on the Environment. From the Detailed Fact Sheet for New Initiatives. Released August 2, 1979. Office of the White House Press Secretary page 24.
17. Gasoline and diesel fuel demands for farms amount to 170 million barrels per year. Since 18% of the current farm land is in corn production and we accounted for the energy inputs to grow that corn in our alcohol energy balance, we need to come up with 18% less gasoline and diesel—140 million barrels per year.

18. The following calculations were made by Charles Washburn based on data from the 1976 Statistical Abstract and from the *Monthly Energy Review:*

$$\frac{\text{Calories}}{\text{Person-year}} = \frac{2500 \text{ Cal}}{\text{day}} \times \frac{365 \text{ days}}{\text{year}} = 912{,}500$$

$$\frac{\text{Calories}}{\text{Car-year}} = \frac{680 \text{ gallons gasoline}}{\text{car-year}} \times \frac{125{,}000 \text{ Btu}}{\text{gal gasoline}} \times \frac{0.252 \text{ Cal (kcal)}}{\text{Btu}}$$

$$= 21{,}420{,}000$$

So you can feed 23.47 people with the number of calories required to feed one car. (These are actually kilocalories, the usual "Calorie" talked about with food.)

This calculation does not, admittedly, take into account the animal feed losses for producing the meat portion of the diet. However, the car value doesn't recognize the 10% conversion loss at the refinery which would be 2.14 million calories.

Also, it may be argued that the distillers' grain protein by-product should "pay for" the energy required to grow the grain, so that no farming energy need be charged to the alcohol produced. However, we need evidence that there is demand for more animal protein than is already being produced.

6

Energy, Water and a Fragile Ecosystem[1]

> My grandfather and grandmother established themselves on the farm that lay nearest to Hope on the north and east. From him I picked up all I know about the life of his generation.
>
> He broke the prairie sod, driving five yoke of straining oxen, stopping every hour or so to hammer the iron plough share to a sharper edge. Some of the grass roots, immemorial, were as thick as his arm. He said it was like ploughing through a heavy woven door-mat.
>
> —Carl Van Doren[2]

The front range of the Rockies defines the western edge of the Great Plains, a dry, ecologically vulnerable, flat land, which tapers eastward, gradually gaining in productivity as the rain shadow loses its grip. At about the 98th meridian, moist air arrives from the Gulf to help irrigate the Midwest, almost completely ignoring the rich western soils of the region.

Not far away from the western edge of this breadbasket, amongst the shadow-causing mountains, is the sunlight of the past buried in coal and shale, and an energy even more ancient than the fossil form—the hot uranium. This is our new National Sacrifice Area, and its proximity to our most tender agricultural ecosystem is ominous. It seems worthwhile to explore the potential consequences of the proximal relationship between the fragile short-grass prairie region and the Rocky Mountain energy area, soon to be exploited. The next wave of U.S. industrialization could easily happen here, perhaps at an unprecedented speed.

The Great Plains have a history of exploitation, a history spelled out in a superb, autobiographical treatment by John Fischer, the former editor of *Harpers.* His recent book, *From the High Plains*, captures much of the essence of Great Plains' economies and ordinary life by looking at the region as an area with a long history of mining ventures.

There were three phases of mining before the twentieth century. The first miners were Indians, who for centuries mined flint for arrows and scrapers, but who had stopped before the Spanish intrusion and, for some unknown reason, before the flint was gone. For about 300 years after the Spanish left the horse, a Comanche economy developed around the American Bison, the buffalo. This was not really a mining economy so much as a harvesting one, but that changed around 1873 when white hunters of hides killed a million buffalo a year for seven or eight years and left the bones to be bleached by the prairie sun. Early settlers sold these bones for fertilizer in the East. Within a decade the buffalo mining was over. Next came the miners of grass, cattlemen who used what should have been the ecological analog of the bison to overgraze land which the migrating Bison barely affected. Drought, followed by an unprecedented blizzard, led to a dramatic collapse of the cattle industry in the 1880's.

The twentieth century mining ventures began with soil mining around World War I. Over fifty million acres of agricultural land in Europe was put out of production, and Americans responded by plowing forty million Great Plains acres. Wheat was planted as a cash crop, and crop rotation was abandoned. Farmers became more active participants in a money economy as they bought newly invented technologies in agriculture, such as threshing machines, for grain production. With their newfound efficiency, they quickly produced a glut on the market. Prices dropped and the response was to plow more land, grow more wheat to pay the debts, and, of course, this further depressed the price. After a few quick rounds of this spiral, which subjected more and more acres to the plow, came the Dust Bowl of the '30's. Some have called it the worst environmental disaster the United States has ever known, and in the opinion of Georg Borgstrom, the well-known Michigan State nutritionist, one of the three great ecological catastrophies since the beginning of agriculture.

By the time of the Dust Bowl, the ephemeral oil and gas boom was already well underway. Almost every pipe driven seemed certain to deliver an abundance of 30¢-per-barrel oil. The history and future of mining this portable liquid fuel is now painfully clear to us all.

The most recent, and what may be the *last*, mining operation of the region is fossil water, much of it from the Ogallala Aquifer (see map at beginning of book). This lens-shaped gravel bed of ancient origin, formerly 300 feet thick and more in places, stretches from western Nebraska to northern Texas, where it supports a twenty-five year old center-pivot economy. Rain recharges the aquifer at one quarter inch per year; farmers and developers are pumping water out at four feet per year![4] Any aerial view shows a non-sustainable agriculture, historically fit only for humid regions, dominating an area in which evaporation is greatly in excess of rainfall.

So far I have described only the mining ventures which (though we often forget) were based on the collective beliefs and values of us all. What a short time ago it was that most of us believed that resources were inexhaustible, that humans can and should conquer nature, and that the good of the individual works automatically for the good of society! Even now, I suppose, millions believe that personal property rights are absolute and that markets can expand indefinitely.

Donald Worster, in his splendid book *Dust Bowl*, has summarized the events and concluded that the dreadful happenings in the southern Great Plains in the 1930's were less the consequence of drought than the consequence of our economic system leaning on these beliefs and values, and finally coming into conflict with a fragile shortgrass prairie ecosystem.

Future of the Great Plains

The most believable scenarios for the future are projections of history. Unfortunately, when mining has been the dominant activity over the landscape, most scenarios which "naturally follow" aren't particularly appealing. Nevertheless, we cannot ignore our history. We will continue to mine water, but there will be a decline in irrigation because of the increased costs of pumping from a lowering water table which will greatly affect the economic base of the region. The following two scenarios, the first negative and the second positive, both begin with the assumption that an "economic hole" will develop in the Great Plains as agricultural income drops.

Scenario One. This scenario arises out of the historical momentum of the past. It is a continuation of a tradition of exploitation of finite and non-renewable resources. A few hundred miles east of the Rocky Mountain coal and oil shale lies the Ogallala Aquifer previously mentioned. North of this fossil water is the Platte River, already impounded by forty small dams, with plans for more.[1] To the south is the Arkansas River, another important regional resource, which is becoming so thoroughly diverted that part of the river bed has become farm ground. Twenty percent of all US irrigated land is in this area. All this water supports a rich, humid-agriculture economy. Some feedlots of the region can accommodate a million cattle at once, or about three million a year. Water supports this, too, for local stockmen no longer have to ship cattle to the real corn belt for finishing but can grow the feed right there. Forty percent of the feedlot cattle are in this region. Consequently, the area now supports numerous packing houses; irrigation equipment and agricultural implement companies line the highways near the towns, and a way of life very different from that before World War II now exists.

Water in the region is progressively becoming too deep to economically lift, even though scarcely half of it is now gone. Few, if any, center pivots are being added, and numerous towns seem certain to experience economic stress.

In 1979, there was a net loss of 14 million acre feet from the aquifer. (By comparison, twelve million acre feet pass Lees Ferry on the Colorado River each year!) Rather than cut back on withdrawals to a level that could sustain irrigation indefinitely, there is now a push on to divert water from the Missouri at St. Joseph and bring it across Kansas in a huge canal and pipeline project. Other plans include diverting water from the Missouri at Fort Randall, South Dakota, and from the White River in Arkansas.

All these plans are nutty, from the energy standpoint alone. Such projects would cost tens of billions of dollars; since the era of cheap liquid fuels is over, they would require electricity from coal or nuclear power plants to run the pumps. As we did in Chapter Two, let us assume that power costs 5¢ per kilowatt-hour and that electrical bills will be half the pumping cost, with capital and maintenance making up the rest. As we did before, let us assume a 75% efficient pump motor, and a cost of $25 to lift each acre-foot of water 188 feet. Let us apply these assumptions to the problem of moving water from St. Joseph, Missouri, which is 850 feet above sea level, to Dodge City, Kansas, at 2480 feet elevation. The height difference is 1630 feet; lifting the water will cost $216 an acre-foot. The cost for replacing the 14 million acre-foot draw down each year amounts to over *$3 billion!* That's for Dodge City. Lubbock, Texas, is at 3190 feet elevation. It would cost $4.35 billion—$311 an acre foot—to bring water there from St. Joseph.

Let us assume that 20 inches of irrigation will give us 140 bu/acre corn. Dryland in Kansas and Nebraska will yield 50 bu/acre of corn equivalent. These 20 inches per acre (1.67 acre feet) of irrigation provides 90 bushels of corn per acre. Therefore, 14 million acre feet of water spread over 8.38 million acres would yield 754 million bushels of corn. If corn sells for $3 per bushel, the gross return would be $2.26 billion for the 14 million acre feet. (If they store the water in the Ogallala, it has to be lifted again, but never mind that.) Before we start subtracting off the ordinary costs for corn production, not to mention the capital costs of the projects, we have a deficit of *¾ billion* dollars for energy and ordinary equipment costs alone!

It all sounds absurd, but the High Plains Governors' Council decided in 1980 to allocate $775,000 for a feasibility study of this project.[3] Fifteen minutes of rough calculations should have been enough to kill the project.

Since industry can afford to pay more for a gallon of water than a farmer can, industry moves in. (The most water-intensive industrial use in California is centralized power plant cooling, and a utility can afford to pay from ten to one hundred times more per gallon than a farmer can.) The Chamber of Commerce leaders in threatened towns will naturally try to attract industry to the region from the East and from Los Angeles and anywhere else they pick up a scent. Lures are water, fifteen-mile-per-hour winds to blow away pollutants, a strong work ethic, dislike of union labor, and close proximity to the newly emerging energy center of the country.

It might seem that once pressure is removed from agriculture to keep the economy alive, area residents can return the plowed land to grass. But what usually happens in a "boom area" is that a "fever" gets generated and land prices accelerate. No one knows exactly why, for relatively little of that land would be needed for housing or for factories. But once high land prices have been paid, there is a strong incentive or pressure, even if the land is purchased as a tax write-off, to make the land pay for itself. Therefore, it becomes wheat ground, not grassland. Unfortunately, the family that has purchased land as a living hearth will have to work it intensely. Suddenly the stage is set for the kind of spiral generated during and after World War I—the one that brought on the Dust Bowl of the 1930's. A nucleus for a massive blow out has been created in a most fragile area. Sure there is stubble mulching now and better conservation practices over all—but extensive acres of dryland farming makes a region vulnerable just the same.

Scenario Two. People of the region realize that a truly sustainable agriculture and rural culture must be based on renewable energy and recyclable materials. Rocky Mountain energy, therefore, is not simply energy to be burned to maintain the status quo, but rather it is seen as *transition energy*. Instead of seeing the wind as an asset for blowing smoke away, it is seen as a potential source of electricity, and thus hydrogen can be produced for vehicular transportation. The area has an abundance of sunlight for heat collectors and photovoltaic cells to make electricity. The water is used no faster than the recharge rate, but is of an ample quantity to turn out renewable technologies. The population of the region experiences modest growth, and the pressure on the land is sufficiently low that much acreage is returned to perennial grassland.

The first scenario leads to "strength through exhaustion" and may happen "naturally." The second scenario leads to a sustainable agriculture and culture, but it will require a deliberate effort to make it happen.

Regional Consciousness: How Soon is Too Soon or Too Late?

Several years ago, when I was at California State University in Sacramento, I directed the Lake Tahoe Environmental Consortium, which involved seven colleges and universities from California and Nevada. Almost everyone involved was interested in preserving what was left of the environmental quality of the

Tahoe Basin. The only problem was that we got there too late, maybe twenty-five years too late. Lake Tahoe sits right where the common boundary of California and Nevada bends. On the California side, second-home development and the usual fast services were a blight. On the Nevada side were the big bucks of the casino owners and operators behind their not-so-hidden values. The formerly pure, clear waters of this mountain gem were now supporting algal growth around much of the edge. What were we to do? We could recommend that excavation be done in such a manner that the soil nutrients were less likely to run into the lake. We could educate children in the schools about the importance of a quality environment. But it was too late to really have much influence because the momentum was already underway.

Perhaps it is not too late to develop a vision of the Great Plains before the momentum for destructive development picks up and perhaps we can promote a future more like the second scenario than the first.

Reference and Notes

1. Much of this chapter first appeared in the Winter, 1980 *Land Report* as a joint effort by Dana, my wife, and me. It was published as a lead off statement for the Land Institute's *Great Plains in Transition* project. I thank Dana for her contribution and for allowing this version to appear here.

2. Carl Van Doren, 1936. *Three Worlds*, Harper and Brothers, N.Y. pp. 25-26.

3. Roger E. Rowlett, May 5, 1980. "Plan Would Send Big Muddy West to Kansas Farms," *Kansas City Times*.

4. James Riser, Sept. 14, 1978. "Intensive Irrigation in U.S. Threatens Water Supplies," *The Des Moines Register*.

7
Agriculture: Tragedy—or Problem with a Solution?

> Well, I've farmed on this land the best I knowed how until it is about to give out just as I am about to give out. Ten years ago I cleared the last cove that was fitten to clear. I don't think a young man could make a living on it now. I don't make near the corn I used to make on the same ground. I can't make enough to feed a horse all the year and have corn for bread too. . . . I don't know. I guess, though, that when all is said our troubles is just because we've lived too long."
>
> —From a case history report[1]

"The essence of dramatic tragedy," said the philosopher Alfred North Whitehead, "lies not in unhappiness, but in the inevitable working of things."[2] As one does a mental survey of the agricultural problem—the failures of success, prophesy, knowledge, eloquence, passion, organization and even stewardship—planet-wide and through time, one is forced to acknowledge the double bind with which we are confronted: without agriculture there will be immediate mass starvation, but with agriculture there will be a continual eroding away of the productive basis of human livelihood. The agricultural problem, viewed in this manner, fits precisely the Whiteheadian definition of dramatic tragedy. Is there no way out of such a trap? Is it our fate that we must merely wait for the unfolding of the drama?

In thinking about these questions, it is difficult to escape considering the Hebrew "Fall of Man" story in Genesis. The Biblical account, you will recall, is one in which an explicit commandment was given by God not to partake of a certain fruit. The first humans disobeyed, and for that error we still pay. This fall is regarded as the basis of the human condition.

There is another "fall of man" story—a modern one—relevant to our situation. A few years ago on the last page of *Life* magazine, I saw a memorable photograph of a near-naked and well-muscled tribesman of Indonesian New Guinea, staring at a parked airplane in a jungle clearing. The caption read something to the effect that the Indonesian government was attempting to bring these savages into the money economy. They had set up a stand on the edge of the jungle and reportedly were doing a brisk business in beer, soda pop and tennis shoes.

We can imagine what must have followed, what the wages of their sin, *their* fall, must have been—decaying teeth, anxiety in a money system, destruction of their social structure. If they were like what we know of most so-called primitive peoples, in spite of a hierarchical structure, they had a much more egalitarian society than industrialized people today.[3]

The New Guinea tribesmen did not receive an explicit commandment to avoid the "goodies of civilization," as Eve, and later Adam, did, who partook of the tree of knowledge. (That is something for academics to ponder.) They were simply *unwittingly accessible* to the worldly items of beer, soda pop and tennis shoes. In Genesis, the sin involves disobedience, an exercise of free will. In the latter version the "original sin" is our very nature, our unwitting accessibility to the material things of the world and to the expedience in our methods of production. Can we really blame ourselves for wanting to be expedient in food production? We have such a long history of agriculture as backbreaking work, and efficient methods helped straighten that bent back.

But let's go back to the New Guinea tribe for a moment and allow a limited extension of our imaginations in order to shed some light on our "fall." Is it not possible that 20 or 30 years after the first beer, soda pop and tennis shoes arrived that some older person of the tribe will remember that things were better before "civilization" arrived literally right out of the air? And isn't it possible that her timidity or apprehension about stepping forward to receive these goodies of civilization may later be translated as a "voice of the Gods" to not partake of these "forbidden fruits of knowledge?" The "fall" of the local culture, which happened *innocently*, is now interpreted after some time as a straight-out refusal to listen to God's order. The resulting problems are from this time forward regarded as the consequence of a sin of disobedience.

It seems natural that a community should try to identify the underlying causes for community-wide problems in order to more creatively work out some lasting solutions. Therefore, we should recognize that it is our very nature we are dealing with, a nature which was not merely harmless in a gathering-hunting setting, but, in fact, was highly adaptive! The chances are that over the millenia any new "resource" (whatever an early tennis shoe analog may have been) introduced into a primitive environment *improved* the chances of survival.

There is another part of our very nature that was adaptive in the paleolithic but which, if exercised now as it was in the past, is decidedly suicidal. That is our tendency to take without thought for the morrow, or, looked at another way, to discount the future. I was impressed with this aspect of our nature several years ago in reading about the Kalihari Bushmen. These modern day "primitives" like to eat a rather large nocturnal mammal called the springhaas. Capturing and killing this animal causes complete destruction of the immediate area. When two hunters discover a burrow, they position themselves at opposite ends and begin to dig furiously with long sticks, throwing dirt in every direction. They have only one thing in mind, to capture and kill this animal.

We could ask for an environmental impact analysis with a paragraph or two on impact mitigation for this area. It isn't necessary, of course, for nature does have redemptive powers, and for that area, redemption is carried out by the reproductive pressure of the springhaas population and by the ecological succession of the vegetation. By the time these or other tribesmen come that way again, the local area likely will have healed. And here is the point: they can afford to take without thought for the morrow, because the long-term ability of the system to support a variety of life and their culture has not been destroyed. The local terrestrial dowry is still there. Even the nutrients tied up in the springhaas will be returned to the general area for recycling. No ecological ethic is necessary so long as the energy put into an area is spent through the arms and legs of the people of the area.

We "civilized people" are a bit better about thinking of our future needs, but we may already have absorbed most of our foresight. During most of our conscious hours even now, we simply take (as we always have) from the environment, without much thought. This pattern of behavior is likely very deep within us and may not be mitigated without internalized pressure from a very strong ethic. Before the fall through agriculture, when our numbers were few, our tools simple, when we were altogether limited in our destructive ability, like the Bushmen, such an ethic was unnecessary.

One of our main troubles, now that our species is out of its natural context, is that we oversimplify our problems and misidentify their roots. The proposed solutions which come pouring forth are not matched to the subtle intricacies of the problem. For example, William Tucker, writing in *The Atlantic Monthly*, suggested that the solution to the erosion of our soils may lie in a revival of organic farming techniques. These techniques did not, of course, save the soils before 1940. But that is a lateral point. He praised the organic farmer, but in a final paragraph he asked if there "are enough people in the country willing to give the time and attention to the soil that is required by organic farming?" The tough-minded producer may reply that we just have to develop the discipline and "shape-up." People born in sin, his unconscious may say, are born undeserving and must exercise the necessary discipline.

The born-in-innocence approach, on the other hand, remembers that nature took good care of us for millions of years before we assumed such a huge role in food production. Since our assumption of more of the burden has steadily undercut nature's chances to provide on a sustainable basis (even though total production on a per acre basis is many times greater), let us look to her again for some clues, for some standards against which we can judge our agricultural practices. If we assume this latter, and I hope more enlightened, myth, isn't it likely we will find ourselves involved in more harmonious agricultural patterns and fewer patterns of destruction? If we make this agriculture less human-dependent and more self-renewing, then the new agriculture, based more on the principles of nature, can afford us a greater opportunity to take without thought for the morrow and still be sustained. As we look to a new agriculture, we cannot nor should we separate our agriculture from our religion or from our ethics. All these are most life-enhancing when they are an inseparable one. So as we move toward a sustainable agriculture, we will necessarily develop an ethic with sustainability at its core.

That we were born in sin is an all-pervasive myth. That we were born in innocence and that our fall is associated with our very nature, which was adaptive in sustainable gathering-hunting ecosystems, may be closer to the truth, but is a long way from becoming an all-pervasive myth. Nevertheless, it is worth trying for. I think this is close to what Henry David Thoreau was talking about over 100 years ago:

> Why should not we enjoy an original relation to the universe?
> Why should not we have a poetry and philosophy of insight and not of tradition, and a religion by revelation to us, and not the history of theirs?

And, we might add, among those works would be a new agriculture closer to our "original relation to the universe." Our old agriculture is rooted in a tradition that is basically ruinous and consequently a tradition from which we must extricate ourselves. We will have to employ the very best practices of conservation and agriculture that are with us today. We will have to continue to support those selfless people and organizations dedicated to the principles and practices of conservation.

But in the longer run, a new agriculture will be necessary. I believe it can and must be done and that when it is underway, it will rank as one of the greatest human ventures of all time, requiring more imagination and truly joyful participation than any of us can now imagine.

We need not feel like the farmer at the beginning of this chapter, helpless actors who must follow an over-written script for a dramatic tragedy which has been 10,000 years in the premiere, and perhaps last, performance for planet earth.

References and Notes

1. Dean Newman and James R. Aswell, 1939. *These are Our Lives*, The University of North Carolina Press, Chapel Hill pp.87–91. This case history report appears as told by the people and written by members of the Federal Writers' Project of the Works Progress Administration in North Carolina, Tennessee, and Georgia.
2. Whitehead, Alfred North as quoted by Garrett Hardin in *Exploring New Ethics for Survival*, 1968, p. 118, Viking Press, New York.
3. James V. Neel, 1970. "Lessons from a 'Primitive' People," *Science* pp. 815–821, 20 Nov.

8
The Religious Dimension

William Blake once commented that "man must and will have some religion." It is hard to deny the truth of Blake's statement though we often skirt this simple reality in our search for solutions to the numerous environmental problems.

Most religions do not emphasize the human-nature relationship even though a search of the written record may produce numerous references to such a relationship. One is tempted to go shopping for a religion that does emphasize more ecologically correct language, milk it for all it's worth, and promote it in order to bring about an appropriate relationship with the environment. If we were to follow this particular approach we would likely abandon anything that smacks of the Judaeo-Christian tradition and latch onto an Eastern religion such as Taoism, Jainism, Buddhism or whatever. But before we strike out on such a tack, there are two things to consider.

First of all, the religions which have emphasized nature are in geographical areas where the environment is in as bad or worse shape than our own. Yi-Fu Tuan pointed this out several years ago, noting that when people are struggling to obtain food and shelter, they often shelve their beliefs and do whatever is necessary to get through.

The second consideration was expressed by the great non-Christian religious leader Gandhi. Gandhi taught that an individual should work out his religious life within the context of his own culture. This enables an individual to take advantage of countless subtleties inextricably interwoven with his or her own culture without having to work at it. Carlos Castenada's Don Juan, the Indian sorcerer of Mexico, explained that his efforts to experience an alternate reality were so extreme because he had no help from the Indian culture, which had been destroyed. It was either carry the load by himself or adopt the culture introduced by the conquistadors. That was the price of his decision to remain "Indian." It may be hard to appreciate the impact of our religious heritage in our increasingly secularized society, but we should look around and see the countless remnants of Judeo-Christian religion which make up our Western heritage.

If we accept our own Judaeo-Christian heritage as the baseline from which an appropriate attitude toward land must spring, what does that heritage have to offer? There are at least two areas of our tradition—besides the usually discussed Biblical passages pertaining to land stewardship—that could promote the cause of land preservation and the livelihood that healthy land supports.

One of the most important and at the same time little discussed traditions is that rules are not locked in hard and fast, that evolution and, eventually, radical change have been possible. A new truth can change the rules. Consider that God, through Moses of course, gave tacit approval to slavery, by giving specific instructions on the treatment of slaves. A long period of time passed from Moses to Lincoln, but finally slavery was abolished, and now this particular scripture is largely ignored. The industrial revolution helped, but justice or a spirit of righteousness or whatever we might call it—a new truth—eventually triumphed over slavery.

There is a new truth before us today which has a great potential value for changing our attitudes toward land and the life on it. In the last 20 years molecular biologists have discovered that the basic structures of all life forms are made of the same 20 amino acids, and except for a few unimportant cases, the genetic code which orders the assembly of these molecular building blocks has one world-wide alphabet. The majority of scientists, at least, accept that all life forms are of the same creation. This really is important, for the Judaeo-Christian myth in Genesis described a *bimodal creation*: God created everything except humans, paused a while, and then went on to create us. That pause has been important in the consciousness of Western society and in Islam. It has contributed to the belief that we are fundamentally different from other forms of life.

Since the new truth says otherwise and since our religious tradition has left room for radical change following new revelation, there is a good chance we can treat the rest of creation better. If we had had a one-creation myth all along, at least in the affluent countries, we might have been less willing to poison the environment "out there" knowing that chemicals harmful to other life forms were harmful to us. In fact, the idea of "out there" might never have existed. Perhaps, like many of the Native Americans, we would have had no concept of such a separation in the first place.

The second part of our religious tradition that could contribute to promoting a healthy and productive biosphere is a fundamental of Christianity. This is the idea of eternal life. It is usually regarded as a reward for an individual's good behavior. This concept has great potential for promoting a sustainable landscape if it can be interpreted to extend beyond the individual to include the eternal life of *Homo sapiens* and all other species. Only by extension of the idea beyond the individual can true transcendence be experienced. It is very much in the Christian tradition, for the emphasis is upon love, love for future generations. This idea need not divide Christians, for what an individual chooses to believe will happen to the individual after earthly death does not necessarily affect the eternal life notion for unborn generations of humans and other species.

These are but two important attributes of the Judaeo-Christian tradition and are mentioned only for the purpose of stressing that a language waits in our cultural depths for an intellectual and spiritual emphasis to give it full bloom.

All of this is simply an argument for embracing an ecological ethic in a spirit of adventure as a part of our religious heritage. None of this involves any abandonment of the religious life of our culture. Appropriately embraced, a spirited and courageous defense of the environment in all of its dimensions can be a part of our own religion. None of this should be taken to imply that the necessary ethical dimensions we will one day embrace have to develop within our traditional churches. The religious life is independent of brick and mortar. For some, association with a congregation is an important spiritual and intellectual experience. However we work through this part of our religious life, it seems important to remember that over 30 years ago Aldo Leopold struggled with the problem of developing a land ethic.

When we consider the evolution of a land ethic we can scarcely avoid thinking about the religious postures which people might take, based on what we know about contemporary religion. How ethical systems arise, and how individuals' values are shaped escapes any formula. There are some wise and religious people who do not conduct their lives according to the orthodoxy of their religious sect, but rather from what we might call a strong, well-integrated *center*. Christianity teaches that each person must work out his or her own salvation. Buddhists try to live a "right livelihood" by seeking the middle way. There are other religious people for whom orthodoxy—religion by formula—is the necessary spiritual warm-up for experiencing the "true essence."

A religious leader may conclude that the religion is starting to "take" when he or she observes that one of the associated ethics is finally internalized. But in the final analysis, it is the individual who must decide whether it has been fully internalized. A young hunter will be aware of his ethical "internalization" when it is no longer a question whether or not to shoot a sitting duck. Orthodoxy may say that to shoot a sitting duck is sinful. Chanting that it should not be done can help internalize the ethic, but this is certainly no guarantee of what the chanter will do when, shotgun in hand, he finds a sitting duck on a lonely pond, the first day of duck season.

We may have it straight, intellectually, what constitutes a sustainable agriculture and culture, but how does the concept of sustainability become such a part of us that our actions and thoughts are appropriately directed? In the environmental movement there are right-wing fundamentalists who have a tight grip and a sharp eye on their do and don't lists, and in so many ways, are altogether unbeautiful and holier-than-thou. At the other end is the liberal environmentalist who can rationalize every excess. One environmental "church" insists that a new kingdom on earth will come when enough individuals live with integrity by minding their own gardens, repairing their own solar collectors and maintaining their own wind electric systems. Others insist that the role of the individual is of less importance than the actions of the group. For them, the collective will—or as Garrett Hardin put it, "mutual coercion, mutually agreed upon"—is the appropriate recipe for promoting a healthy and productive biosphere.

There are arguments on all sides about how to be an effective participant in societal change. Much of the splintering in many religions is the result of the different responses of different groups to the old religious question of how to be in the world but not of the world. Rather than be distraught by this lack of concensus, we should acknowledge it as an important expression of religious freedom and perhaps part of the very basis for pluralism at its best. It forces us to concentrate on the basic considerations of how to be an effective agent for change away from a society which has taken its strength from resource exhaustion and toward a society which uses finite fuels as a transition toward a sunshine future.

In short, I expect the environmentally religious persons in the New Age to be very similar to those who take traditional religion seriously now. But how does this heightened concern become more than religious posturing? We often hear that *more* education is necessary for us to develop an appropriate relationship with the land. Leopold must have been confronted with this easy answer many times, for in *A Sand County Almanac*, he expressed his opinion on the matter in restrained language:

> But is it certain that only the volume of education needs stepping up? Is something lacking in the content as well? It is difficult to give a fair summary of its content in brief form, but as I understand it, the content is substantially this: obey the law, vote right, join some organizations and practice what conservation is profitable on your own land; the government will do the rest.

Leopold believed this formula was too easy to accomplish anything worthwhile because it defined no right or wrong. It called for no sacrifice and implied no change in current values. The formula urged only "enlightened self-interest" and Leopold gave us an example of how far this type of education would take us:

> By 1930 it had become clear to all except the ecologically blind that southwestern Wisconsin's topsoil was slipping seaward. In 1933 the farmers were told that if they would adopt certain remedial practices for five years, the public would donate CCC labor to install them, plus the necessary machinery and materials. The offer was widely accepted, but the practices were widely forgotten when the five-year contract period was up. The farmers continued only those practices that yielded an immediate and visible economic gain for themselves.
>
> This led to the idea that maybe farmers would learn more quickly if they themselves wrote the rules. Accordingly the Wisconsin Legislature in 1937 passed the Soil Conservation District Law. This said to farmers, in effect: *We, the public, will furnish you free technical service and loan*

you specialized machinery, if you will write your own rules for land use. Each county may write its own rules, and these will have the force of law. Nearly all the counties promptly organized to accept the proffered help, but after a decade of operation. *no county has yet written a single rule.* There has been visible progress in such practices as strip-cropping, pasture renovation, and soil liming, but none in fencing woodlots against grazing, and none in excluding plow and cow from steep slopes. The farmers, in short, have selected those remedial practices which were profitable anyhow, and ignored those which were profitable to the community, but not clearly profitable to themselves.

When one asks why no rules have been written, one is told that the community is not yet ready to support them; education must precede rules. But the education actually in progress makes no mention of obligations to land over and above those dictated by self-interest. The net result is that we have more education but less soil, fewer healthy woods, and as many floods as in 1937.

The puzzling aspect of such situations is that the existence of obligations over and above self-interest is taken for granted in such rural community enterprises as the betterment of roads, schools, churches, and baseball teams. Their existence is not taken for granted, nor as yet seriously discussed, in bettering the behavior of the water that falls on the land, or in the preserving of the beauty or diversity of the farm landscape. Land-use ethics are still governed wholly by economic self-interest, just as social ethics were a century ago.

To sum up: we asked the farmer to do what he conveniently could to save his soil, and he has done just that, and only that. The farmer who clears the woods off a 75 percent slope, turns his cows into the clearing, and dumps its rainfall, rocks and soil into the community creek, is still (if otherwise decent) a respected member of society. If he puts lime on his fields and plants his crops to contour, he is still entitled to all the privileges and emoluments of his Soil Conservation District. The District is a beautiful piece of social machinery, but it is coughing along on two cylinders because we have been too timid, and too anxious for quick success, to tell the farmer the true magnitude of his obligations. Obligations have no meaning without conscience, and the problem we face is the extension of the social conscience from people to land.

No important change in ethics was ever accomplished without an internal change in our intellectual emphasis, loyalties, affections, and convictions. The proof that conservation has not yet touched these foundations of conduct lies in the fact that philosophy and religion have not yet heard of it. In our attempt to make conservation easy, we have made it trivial.

Enlightened self-interest not only failed then; it fails today and seems certain to fail tomorrow. The General Accounting Office study in 1976 described its failure in recent times:

County committees generally assigned priorities to the practices for which Federal cost-share funds were to be spent but these priorities

> were frequently not followed. In some cases, practices designated by county committees as high priority or critically needed to control erosion received only a small percentage of the available funds, whereas other practices considered to be production oriented or of a temporary nature were approved by the committees and heavily funded on the basis of popular demand.
>
> For example, during a 5-year period, 52 percent of ACE funds in one county was spent for installing drainage tile in wet cropland and only 27 percent was spent for critically needed erosion control practices, such as terracing and contour stripping. The SCS district conservationist told us that, in most instances, the tiling improved the productivity of the land but provided little erosion control.
>
> In another case, about 80 percent of cost-share funds for a 5-year period was spent to reorganize irrigation systems and only one percent for stripcropping, even though the county and SCS had identified stripcropping as a critically-needed farming practice to reduce wind erosion in areas of the county.

Let us return to Leopold's rhetorical questions of 1948: "but is it certain that only the *volume* of education needs stepping up? Is something lacking in the content as well?" A natural inclination is to suggest that we must be patient, that these things take time. Certainly patience is a virtue, but after 80–100 centuries of a decline in our terrestrial dowry, and at a moment when the decline is at an all-time high, it almost appears as though nature has invented humans for two purposes: to return nutrients to the sea to become sedimentary rock again, and to return carbon dioxide to the atmosphere by burning the fossil fuel. It is ironic that our actions in the name of progress are accelerating the return of this planet to conditions similar to a few billion years ago.

Short-run success in production has only accelerated soil loss. Prophesy has been largely ignored. As hard as the Soil Conservation Service has worked, we still have a severe problem. And we have seen that massive, quality, conservation education through numerous organizations has been tried. We can tick off all these good actions which need to be retained and stepped up, but somehow we still lack the yeast. To repeat Leopold, we are "without an internal change in our intellectual emphasis, loyalties, affections and convictions."

I mentioned early in this chapter the new truth that all life forms are of one creation. Darwin suggested this long ago but the evidence from the molecules is now in. Does not this new truth need the kind of "intellectual emphasis" which Leopold mentioned? The bimodal creation myth did make it easy for us to regard nature as "out there." It helped us adopt a subject-object dualism. It made it easy for us to regard the environment as inherently alien, especially when some part of the environment threatened to eat us or our crops and livestock. Perhaps there was a time when the myth was good enough for reality. But now it seems the myth has the potential for promoting our demise.

There is a final "intellectual emphasis" which could ultimately change our collective "affections and loyalties." This is an unspoken but eventually inescapable lesson of the space program. We have not set foot on the Moon and we likely never will. Perhaps some time in the near future, an objective historian, commissioned to write a brief account of the American space program for a child's encyclopaedia, will accurately describe it as a massive expenditure, involving great technological precision and hundreds of people that allowed a few men surrounded by a terrestrial environment to exist in close proximity to the Moon's surface under the Moon's gravitational pull for a few hours.

The color photographs of our planet taken during the various trips were breath-taking. Partly because of the realization that wherever humans went in space they were obliged to take part of the earth with them, more and more Americans began to see the earth as their permanent home. They began to feel a part of the earth, unwilling and uninterested in taking trips far from the hearth.

I believe that Blake and Gandhi are both correct. Humans must and will have some religion and it is best to work it out in the context of our own cultural and religious heritage. For Western Civilization, that is the Judaeo-Christian heritage, and what it has to offer is rich and filled with hope. It is a tradition that has allowed for an extension of myths so that they may more fully embrace newly perceived realities. The new truth of one creation shows that all species are in it together, that what affects one affects all. Furthermore, there is no environment "out there" consisting of "nothing but" objects. The nonliving world eventually becomes a part of us.

Leopold is correct too. Anything as important as an ethic cannot be written but must evolve in the mind of a thinking community. The ecological church has now and will likely forever have left-wing and right-wing believers in sustainable agriculture and culture. But regardless of the religious posturing, it will be necessary to give some intellectual emphasis to the new truths and eventually to influence the loyalties and affections of us all.

9
The Farm as Hearth—Or the Farm as Food Factory

Dinuba and Arvin: A Lesson on the Beauty of Small

When the land begins to be regarded, not as the primary source of wealth, but as the plaything of gentlemen already rich, the economy of the country is in questionable, if not dangerous condition. England, to be sure has survived in spite of that attitude; but only by becoming the workshop of the world.

—Gerald W. Johnson, 1939[1]

Dinuba and Arvin are two separate communities approximately 120 miles apart in the great Central Valley of California. The apparent ecological constraints on each community are quite similar, but here the similarity ends. In the mid-1940's, Dinuba, to the north, was a small-farm community; Arvin was surrounded by large farm holdings. In 1950, Dinuba had 4940 people and Arvin 5007.

Beginning in 1944, Professor Walter Goldschmidt[2] and his associates at the University of California began to gather extensive sociological data, the interpretation of which has had a resounding impact on those interested in the promotion of a healthful and productive rural life. Their results have much to say about the character and quality of rural community life as the consequence of size. The most simplified conclusion is that "small is more beautiful than large."

The small-farm community, Dinuba, at the time of the study had 62 separate businesses; Arvin only 35. Over a 12-month period during the study, Dinuba experienced more than $4.38 million dollars worth of retail trade while Arvin managed a retail volume of around $2.53 million. Dinubans spent three times as much for household items and building equipment as did members of the Arvin community; and, furthermore, the dollar literally went farther in Dinuba, for it supported 20 percent more people. Using usually-accepted indicators for measuring standard of living, life was better in Dinuba than Arvin. This manifested itself in several ways. Only 30 percent of the employable work force in Arvin was independently employed either as businessmen or farmers. Dinuba, on the other hand, had more than 50% of its comparable force in such positions. Less than a third of the Dinuba people were landless and worked as agricultural laborers; almost two-thirds of the Arvin community breadwinners were landless laborers. The list of differences between the two communities consistently reveals positives associated with the small-farm community and, relatively speaking, negatives associated with its opposite. People in Dinuba enjoyed more paved streets, more sidewalks, more garbage disposal equipment and sewage disposal conduits. They maintained four elementary schools and one high school, while Arvin supported only one elementary school. Dinuba had three parks or playgrounds; Arvin had only one—loaned to it by the corporation. Booster-type civic organizations were twice as abundant in Dinuba as in Arvin. Dinuba benefitted from two newspapers, each with many times the total news space as the one Arvin paper. Finally, Dinuba had twice as many churches.

The Dinuba-Arvin story is an account which we proponents of decentralization like to hear. We have every right to believe that the farms in the Dinuba community approach the hearth model. They certainly do in size. Even though skeptics might argue that the differences in the quality of life can't be all that great in the two towns, it is obvious that an egalitarian sharing of regional wealth is more seriously denied in the town surrounded by large holdings. The U.S. Department of Agriculture tried to keep the lid on the study; and it was discredited from several quarters, probably because, as Senator Gaylord Nelson has said, certain "business interests well knew the social consequences of their modes of production."[2]

The problems of inequitable ownership nationwide are even worse now than they were in 1944. This has had a tremendous impact on the cash returns from agricultural efforts. About half our farms (1.25 million) are responsible for only three percent of all farm products. Twenty-five percent, on the other hand, are responsible for 85 percent of the products.[3] Seventy percent *grossed* under $20,000 in 1978 and accounted for a scant 11% of total cash receipts from farming![4] If we stop right here we have most of the story of the inequities in American agriculture, but the story is longer and it gets worse. In fact, inequities have become more acute in recent years.

From 1971–1977, the capital gains on the physical assets of farmers, in fixed dollars, outstripped net income from farming two-and-a-half times over.[5] This is sobering. This only happened because, increasingly, land has become an item of speculation and a hedge against inflation, mostly for people who have more money than they need. Aside from the tax advantages which come from buying land, those with big money can afford to sit it out—that is take the small income on the large capital gain, and let inflated dollars buy the farm. Contrary to the rhetoric of the Nixon administration, which said that everyone loses from inflation, some people *do* gain. Speculators now own increasingly large numbers of acres. The consequence is that in the short run when land is priced above its earning power, the small farmer, or at least the farmer with little money and few assets, is driven out.

The bottom line of this problem is that there is little chance to re-establish the traditional structure of agriculture without drastic and bitter problems. We already have the unhappy situation that three percent of the people own 55 percent of all land and 95 percent of all private land holdings.[6] Given the rules of our economic system, one can scarcely blame investors who seek to guard themselves from the uncertainties of our economy. But is all this growing largeness of scale in agriculture due to inevitable consequences which have resulted from the nature of our economic system?

Larger and Fewer Farms: Inevitable Consequence or on Purpose?

The last 50 years saw our country lose four million farmers. In one year alone, 37,000 farmers went out of business. As of 1979 there were a scant two and a third million farms in the U.S. This group is increasingly powerless, politically, and they have blamed the U.S.D.A., agri-businesses, the farm press and the land-grant colleges. In the minds of most, however, the economic problems which led to a de-peopleing of our countryside were the consequence of "natural" circumstances. Examples of "natural" problems include such well-rehearsed conventional wisdom as "farm units were too small," or "land was unproductive" and "some farmers were simply poor managers," or "too many well-intended farm programs didn't work," and finally, "America's cheap food policy was implemented at the expense of the food producers."

All of these standard reasons have some merit, and for most of my thinking years I accepted them as the appropriate wisdom, if taken in the right mix. Since boyhood I have heard "extremists" at gatherings of farmers, whether at a sale barn or at a farmers' meeting, angrily blame the demise of the family farm on some conspiracy. "Mature judgment" required that I dismiss most of this talk as a form of paranoia lacking substance or fact. After all, no sufficiently large group of corporate managers could possibly get together to orchestrate such a cruel, un-American and altogether unhealthy social change, either for the evil purpose of lining the pockets of a managerial few or—and many people probably think this—"for the good of the country."

However, documentation about just such a policy change is now being marshalled and could well provide the basis for a strong grass roots movement for massive land reform legislation.

Mark Ritchie, author of a small booklet entitled "The Loss of Our Family Farms," has done an excellent job of extricating and synthesizing material from various reports published by the Committee for Economic Development.[7] CED is one group of men who have advocated the removal of people from farms.

Organized during World War II, most of the committee were business leaders concerned with the "mass unemployment" sure to result after the war effort brought us victory. In all fairness, we must remember that fresh in the minds of these captains of industry was the depression of the 1930's. The 30's had fostered a great deal of social and political unrest; large numbers of people called our entire system of capitalism into question. Those who stood to lose the most, if this system should crumble, sought to avoid the problems. It was in such a psychological ambience that the Committee for Economic Development was formed. Part of the group consisted of corporation presidents known for their strong business sense, "experience in analyzing issues" and their promise for "developing recommendations to solve the economic problems that constantly arise in a dynamic and democratic society."[8]

Another major group of participants included several university presidents. Early on, they explained that "through this business-academic partnership, CED endeavors to develop policy statements and other research products that commend themselves as guides to public and business policy: for use as texts in college economic and political science courses and in management training courses, for consideration and discussion by newspaper and magazine editors, columnists and commentators, and for distribution abroad to promote better understanding of the American economic system."[8] They certainly had faith in the role of education for they essentially advocated an information "blitz" at several levels.

The Committee for Economic Development suggested that the main problem was the persistent excess of resources in agriculture—particularly *labor* relative to the new farm technologies.[8] The men at CED had identified part of the problem as too many farm workers. They seemed puzzled that these people were reluctant to leave the farm. There must be something awfully compelling about the farm. However, these intellectuals and businessmen reckoned that it was the support of prices that had deterred the movement out of agriculture. We should not forget that these were men who were accustomed to making things happen. They wanted the exodus from the farm to be "large scale, vigorous, and thorough-going." They proposed that the farm labor force, five years into the future, be no more than two thirds as large as its then current size of 5.5 million. "The program," they said, "would involve moving off the farm about two million of the present farm labor force, plus a number equal to the large part of the new entrants who would otherwise join the farm labor force in five years."[8] In other words, get two million off the farm and keep their sons from staying on the farm.

They had a plan for accomplishing this. Simply lower the guarantee on agricultural products by lowering parity, and economics would do the rest.[8] The language of the committee is less blunt, which may indicate that even some of them were sickened by the thought. But they had a mission, a job to do, and they explained that the basic adjustment required to solve the farm problem, adjustment of resources (mostly people?) used to produce farm goods could not be expected to take place unless the price system were permitted to "signal to the farmers."[8]

CED recommended that price supports for wheat, cotton, rice, food grains and related crops be reduced immediately.[8]

They insisted that the importance of such price adjustments should not be underestimated. The lower price levels would discourage further commitments of new productive resources to those crops unless they appeared profitable at lower prices. On this point the committee was most emphatic. They stressed that:

> For several reasons it is important that price supports be moved to levels that, if wrong, will be low rather than high. Second, new resources (especially people) should be discouraged from entering agriculture, at least during the adjustment period, and the rate of entry in the longer run should not be excessive.[8]

These "leaders" apparently recognized that the sons of farmers would think that, "if it is too tough for Dad to make a living here, why should I try?" The result: two million farmers and their families were displaced!

There were other effects which this blue ribbon group anticipated. Lower prices meant more exports, and since many of them doubtlessly operated as middle men in at least some of their enterprises, the price of the item was of less importance than the fact that they were there when the money changed hands.

Ritchie summarized how these powerful people must have viewed the primary benefits of their recommendations: (1) increased return on corporate investment in agriculture, (2) over two million farmers and families entering the urban labor pool, which would tend to depress wages, and (3) lower prices on agricultural products, which would increase foreign trade and provide cheaper raw materials for domestic food and fiber processors.

There are probably many respected agricultural economists—including some who believe in the sustainability (rather than production-only) paradigm—who are convinced that CED had little if any effect on farm policy. They might be right in the most direct sense. But what is more important is that psychological ambience I mentioned earlier—the collective mind-set of those with direct money interests in the food system. The communication throughout this establishment is complete, even though most members of the network may never talk to one another. We have long known that the connections among the various strong corporate interests have included our top universities and Washington policy makers. Many of these people are part of a high level "good-ol'-boy" system loaded with "mature judgement." They share a *presumption* that they know what is good for the country—and, for that matter, the world—and from that assurance they wield their power to influence policy, usually subtly, sometimes not.

Of course many of them are altruists, often toward poor people in foreign lands who need what they deal in. And when their wisdom is sought they understandably advocate the kind of policy which is both familiar and has worked—for them. "If it is right for us," they seem to say, "it must be right for the rest of the world."

Let's look at a concrete expression of this attitude. Remember that CED wanted their policy projected into the classrooms. In the college text *Economic Development*, by Gerald Meier and Robert Baldwin, is a description which reflects the "ol' boy" presumptions. In a chapter entitled "General Requirements for Development," the authors describe how economic criteria of investment may not be sufficient to bring about the necessary changes, and that some non-economical actions may be to "invest in projects that break up village life by drawing people to centers of employment away from the village."[9] The authors continue by stating that "new wants, new motivations, new ways of production, new institutions need to be created if national income is to rise more rapidly. Where there are religious obstacles to modern economic progress, the religion may have to be taken less seriously or its character altered." Fundamentally the backward peoples must recognize that men can master nature; they must be motivated towards economic achievement; they must acquire the means of accomplishing these objectives; and these objectives must become part of the society's value structure."[9]

Out of the same cloth came the recommendation for a systematic attempt to take power away from farmers and force them into town to re-allocate the labor pool—for the good of the farmers, of course, and for the good of the country.

It may be argued that it was a good idea to de-populate the countryside, to destroy thousands of communities, rural churches, schools and most importantly, rural values, but we cannot seriously argue that it happened "just naturally."

It appears that the plan has worked, for the decentralized rural life has mostly vanished. It is doubtlessly true that much of rural life was a drag and that the city offered new hope, new beginnings, more stimulation and a kind of sophistication. But as bad as it may be, the people did not move away in large numbers until forced to by the economic squeeze.

Farm population had already been in decline and it is hard to know exactly how much influence the "committee" had over government policy and how much was due to a momentum already under way. We may not be able to trace all the tethers which led from this committee to people in the various administrations, but it would be naive to believe that the connections were few and the "committee's" influence weak.

The Relation to the Land Americans Have Historically Wanted

The Jeffersonian ideal of a democracy based on agrarian values and enlightenment was alive and well as recently as 1862 when the Homestead Act established the system of small farms during the era of great expansion. A settler, for a small fee, could acquire 160 acres free and clear if he would live on the homestead for five years. People with the will and strong bodies could own land. With westward expansion into more arid areas, 160 acres wasn't practicable, and statutes were drawn which would encourage water resource development by allowing larger farms. By 1900 it was clear that these laws had failed to provide enough irrigation water or enough people on the land, and that our policy makers' good intentions weren't enough. As historian Arthur B. Darling concluded, "In spite of every measure yet devised, the nation's resources in land were accumulating in large tracts owned by a few wealthy individuals and corporations."

Let us look at another noble effort from our history. The Homestead Act was scarcely five years old in 1867 when Congress passed the Morrill Act for the explicit purpose of training farmers and artisans. Twenty-five years later the Hatch Act added research, and fifty years later the Smith-Lever Act provided for Agriculture Extension. The purpose of all three of these acts was in every way noble—the "democratization of knowledge,"[10] as Dr. Harold F. Breimyer of the University of Missouri describes it.

The National Reclamation Act of 1902 was the next major land legislation drawn to ensure people's presence on the land. This statute came forty years after the Homestead Act and declared:

> No right to the use of water for land in private ownership shall be sold for a tract exceeding 160 acres to any one landowner, and no such sale shall be made to any landowner unless he be an actual bona fide resident on such land, or occupant thereof residing in the neighborhood of said land.

Clearly, Congress was trying to close the loop holes which developed in violation of the spirit of the original Homestead Act.

The law makers, acting in response to their constituents, have professed that the rural life is important to American society. We have long recognized the importance of being rooted in the land and at least since the Land Grant College system was established, the importance of the necessary education to go along with this relationship to the earth.

In the meantime, the land has fallen into fewer and fewer hands and increasingly the Land Grant Universities seem to have betrayed their historical commitment to the "democratization of knowledge." This did not happen just recently. Over thirty years ago, Ralph Borsodi lashed out at the public institutions in our agricultural system as "blind leaders of the blind," and accused them of "deliberately commercializing and industrializing agriculture: of subordinating the real interests of agriculture to that of the fertilizer industry, the canning industry, the agricultural implement industry, the seed industry, the milk-distributing industry, the meat packing industry, the automotive and petroleum industries, and all the other industries and interests which prosper upon a commercialized agriculture."[11] He accused them of "teaching the rape of the earth and the destruction of our priceless heritage of land . . . of impoverishing our rural communities, wiping out our rural schools, closing our rural churches, destroying our rural culture, and depopulating the countryside upon which all these are dependent."

If What We Wanted Is Not What We Got, Where Did We Take The Wrong Turn?

We can deal with this question on two levels—the philosophical level and by looking at explicit examples. One is not more important than the other.

The great American writer, Thomas Wolfe, in the "Credo" of *You Can't Go Home Again,* describes what others have known through the ages. Wolfe thought the true discovery of America is before us and what he calls the "enemy" is before us as well.

> . . . the enemy is single selfishness and compulsive greed. I do not think the enemy was born yesterday, or that he grew to manhood forty years ago, or that he suffered sickness and collapse in 1929, or that we began without the enemy, and that our vision faltered, that we lost the way, and suddenly were in his camp. I think the enemy is old as Time, and evil as Hell, and that he has been here with us from the beginning. I think he stole our earth from us, destroyed our wealth, and ravaged and despoiled our land. I think he took our people and enslaved them, that he polluted the fountains of our life, took unto himself the rarest treasures of our own possession, took our bread and left us with a crust, and, not content, for the nature of the enemy is insatiate—tried finally to take from us the crust.
>
> I think the enemy comes to us with the face of innocence and says to us:
>
> "I am your friend."

At the second level, we need to ask how the dreams of our ancestors for promoting the decentralized rural life have gone afoul. How and where did the "enemy" get into our institutional structure? The "enemy" was present when the partnership between public-supported research and agri-business began, when agri-business approached our institutions and said, in Wolfe's language, "I am your friend . . . See, I am one of you . . . Behold how rich and powerful I am—all because I am one of you—shaped in your way of life, of thinking, of accomplishment." And so numerous corporations backed by power, posing as friends (indeed, believing themselves to be friends), but acting with special interests, developed partnerships with the public-supported research establishments by partially funding hundreds of research projects, and gained up to a five-fold leverage on the publicly-funded staff and facilities. In addition, faculty and researchers did private consulting, and eventually the commercialized interests and the public interests came into conflict.

This is probably the most written-about socio-political problem in modern agriculture. That important subject cannot be adequately discussed here, given the scope and purpose of this book. Wendell Berry, in his marvelous book, *The Unsettling of America,* has written eloquently and with great passion on this subject and I hope any reader who has not read his book will.

Part of the problem is institutional, but most has to do with the ordinary human limitations Wolfe described. When the extension arm was added to the Land Grant College System, the connection of the academic specialist to the needs of the farmer was one of the most important decisions in the history of agriculture. Wolfe would undoubtedly regard it as the true American character shining through. I believe I saw it during my early childhood in the kindly extension man from Kansas State, who in the midst of the depression, and through the war years, advised our family on practices for our farm in northeast Kansas. My parents had six children to support on that small farm at the height of the depression and, as I have heard them say, many times over, almost prayerfully, they "never went on relief." This pipe-smoking gentleman, (I can see him yet) went from farm to farm in our Kansas River Valley, and I am sure in the adjacent uplands, with helpful information, "from the college." This image was enough to carry me to adulthood where years later, I would take an advanced degree from the College of Agriculture of a land-grant university in another state.

But something has gone wrong. I say this reluctantly for many of my best friends and closest colleagues are associated with this system and I have had and continue to have many positive experiences with it. I know there is still some of that true American Character there that Wolfe described. I point out this history, for though much of the relationship with these agricultural institutions has gone sour, in the larger picture, any quarrel I have with this system as it now stands, is truly a "lover's quarrel." Furthermore, many of my friends in this system feel much as I do. I should dedicate this chapter to them for it is their truly American values (again, in the sense Wolfe meant it) that give me hope that a new emphasis will one day be possible.

But I promised to deal with some specifics of how the dreams of our ancestors for decentralization and a good rural life have gone afoul. There were at least three problems or categories of problems which were to develop eventually, limiting the effectiveness of this beautiful formula. All good things do bear the seeds of their own destruction, and as we place ourselves on the correction course, we should recognize some of the shortcomings in the relationship of education to agriculture. These three have been outlined by Dr. Harold Breimyer: they are academic careerism, clientism, and the slot-machine approach to dispensing knowledge.[10]

The first problem results from the very important need for professors to maintain a certain amount of academic independence. Unfortunately, tenure and promotion, both very necessary in our citadels of knowledge, are achieved the quickest and easiest by publications built on an old bibliography and research. The problem begins to develop when young Ph.D.'s need tenure. The behavior doesn't automatically stop after tenure, for there is always the desire for promotion. Any practical question from a farmer which does not pertain to the academic careerist's own area of expertise may be seen as a bother or a distraction, even though he or she may be the best qualified to deal with the problem.

Another factor in our society has simplified the problems for institutions, at least in the short run. It has to do with cheap oil. We can all remember that a few short years ago the oil companies had such a glut of their product they went shopping for ways to use it; we all remember the glasses and trinkets service stations used to give away. Three or four decades ago, they started walking into the public-supported research establishments with some research money, and one thing that came out for agriculture was our current oil-based pest control program. These crude but all-sweeping methods of pest control, in turn, allowed numerous researchers to pursue more esoteric scientific questions.

We lost a lot of time when researchers could have been developing truly sophisticated methods of pest control. Unfortunately, lost time is not the only problem, for now we anticipate several potential problems associated with many of the oil based chemicals. It should not be taken too lightly when someone charges that the U.S. Department of Agriculture and the Land Grant University system have become subsidiaries of the petro-chemical companies. I do want to emphasize that academic independence is necessary—but we need to remember that it can be abused by the careerist who sees his training as a "passport to privilege." The purpose of such independence is to allow the creative mind to work in behalf of society.

A second major problem has to do with the Extension Service, inherently the most vulnerable of public organizations, especially if it lacks a vision of its role in society. Clientism is a condition associated with public agencies in which the public servant waits on whoever shows up and in about the order that they do show up. Without a vision or a mission or a policy of purpose, moneyed interests soon tie up a disproportionate share of time and resources.

A third problem is what Breimyer calls the "slot-machine" dispensing of knowledge. If I have cabbage worms on the broccoli, I can call the extension agent and he will send me a brochure telling how to spray. It is a way of dealing with the needs of people who aren't very powerful.

Powerful interests, not the least of which has been the Farm Bureau, have sought to exercise control over one or more of our public institutions of agriculture.[12] Masters' theses are turned out on packaging, instant mixes, micro-wave cooking of meats. The all-pervasiveness of the packaged food industry as well as the suppliers of products which enhance production at the expense of a despoiled land is a well known fact. The problem was not born on such and such a date. I believe, as Wolfe says, that the problem "has been here with us from the beginning and comes to us with the face of innocence."

It is up to all of us to rededicate ourselves to the old vision of the farm as hearth and to be vigilant and to opt for nothing less than that which leads to a sustainable agriculture.

Food Factory, No!
Hearth, Yes!

Few who have seriously thought of the long term future of food in America doubt for a moment that farming as a way of life needs to be promoted, not for the purpose of providing museum pieces for city dwellers, but because we need stewards on the land. Even the town and urban population, in the not too distant future, will have to look to the land reverently, as the source of their sustenance and health. By then it should have become increasingly clear that stewardship based on economics alone won't do, for if farming continues as a business proposition only, the land is doomed. Eventually short-run economics will dictate the patterns of use. Most small farmers would have it otherwise. Most of them would prefer the farm to be a home or a hearth, a place to live and raise a family.

In the coming age, land tenure rather than land price should be the watchwords. *If air, space, sun and food really are more important than dollar values and development potential, we must remove land as an item for speculation*. But it can't stop here. One of the logical dangers is that more and more land could still fall into the hands of corporations, and a monopoly with all the attendant evils will result. We must figure out ways to guard against this. In this respect, it might be important to re-read the history of the great French Revolution or the revolt against the Church in South America during the last century.

As we begin to promote the rural life, we should be realistic about the potential of the various land types; here are a couple of considerations to think about. (We are much too hopeful many times as we see society's evils fall away in our visions of decentralization.) Let us assume that forty acres seems like an appropriately-sized piece for a family. If we tried to put every family of four on 40 acres, or one person on every ten acres, given our population of some 225 million, we would need about 2.25 *billion* acres. We have roughly 400 million acres of agricultural land, an additional 100 million acres of marginal land and about 800 million acres of range land about equally divided between private and government ownership. In other words, we have 1.2–1.3 billion acres of food producing land in the country. We would need to nearly double our total acreage just to accommodate each family on 40 acres.

But this relationship between people and acreage is meaningless in many respects. When only one person is on the moon, the people/land area ratio is the most favorable of any place in the solar system. But at that moment, the moon is also the most seriously over-populated piece of real estate in the solar system, since the carrying capacity of the moon is exactly zero. Population density per square mile is meaningless. One family on a square mile of mountainous plateau in Utah may experience less room than 50 families on a square mile of river delta in Louisiana.

When we speak of decentralizing and even regional *semi*-self-sufficiency, we must ultimately consider who is going to be shooed or lured where. Let us assume it is a simple lure like "five acres and independence." Where are these acres? How many frost-free days? What will the land grow? Who is the person or family that will grow it? What is the slope of the land, the rainfall, hours of sunlight, usable fertility, and what is left of the initial dowry? Is there a market for the produce? What machine equipment will be used? How much does the land cost? Is it paid for? If not, what is the interest? This is the first battery of questions. As staggering as these questions are, they are the easiest to get answers to—and they may still not be the most important. The society should try to figure out the energy and materials cost for laying down the minimal infrastructure necessary for decentralization. It is likely to be substantial. A society bent on decentralization should probably have a National Policy for Agriculture and Rights of Tenure spelled out early on. Like Thoreau, we need to ask, "What do we want of this land?" and I hope that like Thoreau, we can conclude that we want to "rear our lives to an undreamed-of height and meet the expectations of the land."

But again, none of this can come about without the will to make it happen. Paul S. Taylor, Emeritus Professor at the University of California at Berkeley, who has studied public policy and the shaping of rural society, has concluded that Americans have passed ample legislation to support the family farm, but that administrators have failed to carry out the purpose of the acts.[13] Taylor has concluded that political as well as economic realities now demand action from Congress that will enforce a policy *favoring* the family farm. Without such a policy, Taylor concludes, family farming will be lost as a way of life.

Part of such a policy should be the recognition that land is the birthright of the people. As such, both control and ownership would be widely distributed so as to diffuse economic and political power. This is very important for there seems to be an emerging law of human ecology that people in possession of, or very close to, power are unlikely to be close to the heart of a problem. To be close to the ecological problems of agriculture, the people who live and work on the farms should either own the land or be participants in a land trust system in which everyone's first interest is the conservation of healthy land and water. The bottom line cannot be profit. The land should not be owned by large corporations or wealthy absentee owners but if it is, policy measures should ensure that there is compliance to promote and achieve the best soil and water conservation possible.

In the last forty years we have encouraged absentee ownership, and a rural exodus has resulted. New government policies will be needed to reverse that trend. The language should be unequivocal in this regard, as tax laws that now encourage ownership by speculators or absentee corporations with nonfarm interests, preferential treatment for capital gains, depletion allowances and underassessment, will have to be repealed. Numerous human social services, such as education and health care, can be supported by revenues derived from progressive property taxes on the unearned increase in land value. Services of the U.S. Department of Agriculture and the Land Grant Universities should be offered exclusively to family farmers and consumers.

But this won't be enough. The young farm family will need capital to get started. Surely it is within our tradition to provide the opportunity for young, able and willing people to enter farming, as through low interest loans or reduced taxes for a certain period.

In the meantime, people everywhere can begin to take advantage of the corporate model of ownership by investing their money in land trusts. As small farmers are forced out, the trusts can buy the land and give the priority to those who have owned the land, asking them to stay and even offering them the opportunity to buy into the trust. The task of the owners would be to draw up the rules on how the farm is to be managed, to ensure that the bottom line is not profit but healthful conservation. There will be years when yields are low and the temptation will be high to increase productivity at the expense of the future.

It is a humble beginning but it is a possible beginning. It is certainly a way much of the hollow savings and investments can be transferred to something real and lasting.

There are still hundreds of opportunities in agriculture though the number of options is decreasing fast. Historians tell us the frontier closed just before the turn of the century. This is true enough, but in one sense it started closing when the first tobacco was planted at Jamestown as an export cash crop. The psychology of Jamestown has shaped our laws. So did the father of our country; Washington had huge holdings. So did the railroads, the greatest land holder of all. It is important to remember that our opportunities for carrying out visions of a wholesome, decentralized society are severely limited by the heritage of our ancestors who had their own "expectations of the land."

There has always been a minority in our country who make a practice of meeting the land's expectations first. Can we ever expect this number to increase? Is there any reason to believe that most of the countryside will ever have the chance to respond to the touch of a people who believe in and cherish such a relationship?

My answer is a cautious or qualified "yes."

An increasing number of people are finally recognizing that agriculture itself is an ecological problem outranking industrial pollution. They are not in the ranks of those who have only worried about farm chemicals, such as pesticides and fertilizers, as an outgrowth of their worry about industrial pollution.

Those who are concerned about agriculture are asking more astute questions. The questions to which these growing numbers are seeking answers are getting better all the time. Even so, we are talking about a journey—a very long journey which our species may not have time to finish.

But then, even if it doesn't, is there anything better to do?

References and Notes

1. George W. Johnson, 1939. *America's Silver Age,* Harper and Brothers, NY. pp. 43.
2. Walter R. Goldschmidt, 1978. *As You Sow: Three Studies in the Social Consequences of Agribusiness*. Allanheld, Osmun & Co. Montclair, N.J.
3. See Farm Income Statistics, USDA, ESCS, Statistical Bulletin 609, July 1978 and also see Harold F. Breimyer's paper "Who Controls Agriculture in the Great Plains?" Univ. of Missouri-Columbia, Ag. Econ. Paper No. 1979-27, p.2.
4. From a draft statement issued by the Catholic Bishops of Regions VII, VIII, IX entitled "Strangers and Guests: Toward community in the hearthland," 1979.
5. Harold F. Breimyer, 1978. "The Issue of Foreign Purchase of U.S. Farmland: A Reflection on Principles." Agricultural Economics, Univ. of Missouri-Columbia.
6. Peter Meyer, Jan. 1979. "Land Rush: A survey of America's land." *Harpers*.
7. "The Loss of our Family Farms" can be ordered by writing Mark Ritchie, 824 Shotwell, San Francisco, CA 94110.
8. "An Adaptive Program for Agriculture," A statement on National Policy by the research and policy committee of the Committee for Economic Development, 1962, p. 2.
9. Gerald Meier and Robert Baldwin, 1957, *Economic Development*, John Wiley and Sons, Inc.
10. Harold F. Breimyer, 1978. "Outreach Programs of the Land Grant Universities: Which Publics Should They Serve?" Keynote address at Kansas State University for conference with same title and address. pp. 5–17.
11. Ralph Borsodi, 1948. "The Case Against Farming as a Big Business," Vol. VI:4 *The Land,* Vol. VI:4 446–451.
12. William J. Block, 1960. "The Separation of the Farm Bureau and the Extension Service," Vol. 47, The Univ. of Illinois Press, Urbana.
13. Paul F. Taylor, 1975. "Public Policy and the Shaping of Rural Society." *South Dakota Law Review,* Vol. 20.

10
The Search for a Sustainable Agriculture: A Bio-Technical Fix

The grass is rich and matted, you cannot see the soil. It holds the rain and the mist, and they seep into the ground, feeding the streams. . . . It is well-tended, and not too many cattle feed upon it; not too many fires burn it, laying bare the soil. Stand unshod upon it, for the ground is holy, being as it came from the Creator. Keep it, guard it, care for it, for it keeps men, guards men, cares for men. Destroy it and man is destroyed.

—Alan Paton[1]

Not content with the authority of either former or present day husbandmen, we must hand down our own experiences and set ourselves to experiments as yet untried.

—Columella, 1st century A.D.[2]

We have seen that nowhere is the ancient and long-discussed split between humans and nature more dramatic than in the manner in which land is covered by vegetation. To maintain the "ever-normal" granary, the agricultural human's pull historically has been toward the monoculture of annuals. Nature's pull is toward a polyculture of perennials. This is not to say that we humans exclude perennials from our agricultural endeavors, just as nature does not exclude the annual plant as part of her strategy to keep vegetation on the ground. Certainly the numerous nut and citrus trees, grapes and berries (be they blue, black, rasp or straw), along with other perennial plants, are important to our species. As for nature, no naturalist need remind us that her annuals are widely dispersed in natural ecosystems.

The main purpose of this chapter is to consider the implications of these opposite tendencies, with an eye to the serious work involved in healing the split. Nature is at once uncompromising and forgiving, but we do not precisely know the degree of her compromise nor the extent of her forgiveness. I frankly doubt that we ever will. But we can say with a rather high degree of certainty that if we are to heal the split, it is the human agricultural system which must grow more toward the ways of nature rather than the other way around.

Nature rewards enterprise on a limited scale. A weedy annual is enterprising. Not only will it cover bare ground quickly, but it will yield an excess of potential energy besides. This is probably the reason our most important crops, such as corn and wheat, arose as weedy annuals. A small amount of annual vegetative biomass promotes the production and survival of a rather large number of seeds during a growing season. This is usually assured by one of three things or even a combination of all three: (1) the storage of plenty of food in the seed, (2) the set on of many seeds and (3) the ability to colonize a disturbed area. Many perennials may have these three characteristics, but it is less critical for them to come through in a particular season, for there is always another year. For that matter, there is always another year for many annuals too, as their seed will remain viable for more than one year. But overall the colonizing annual has had to rely on enterprise. The ancestors of our current crops may well have been camp followers, colonizers of the disturbed ground around the campsite. They were obvious candidates for selection by humans because of their availability and their inherent ability to produce an excess of potential energy. They are the enterprisers of the higher plants.

We don't know whether the early agriculturists were faced with famine or not. But when they began to plant annuals in fields, they were beginning to reward this enterprise. The monoculture of annuals, the enslavement of enterprising species, was a big new thing in the history of the earth. The face of the earth was changed.

By and large, the patient earth has rewarded patient ecosystems, but it would seem that enterprise has always been rewarded too, though on a very limited scale. It would seem to be a good strategy for an ecosystem to have such species present, for these quick colonizers could rapidly cover the ground made naked by a migrating buffalo which had wallowed and dusted himself, or by an excessive flood or an insistent wind. The ecological capital which had been sucked from parent rock material or stolen from the air could be retained to promote more life for future generations of all species in the system.

The selection of enterprising plant species has rewarded all humans bent on enterprise in food production.

But there is the second consideration already mentioned in Chapter 8—the likelihood that we have a psyche predisposed to take from the environment with little thought for the future, especially when the connection between the product and the source is separated by numerous links. It is the combination of these two psychological characteristics, enterprise and taking with little thought for the future, which has resulted in a rub yet to be reckoned with in the four hundred generations since humanity started keeping track of seed time and harvest. The problem is this: to maintain any system, agricultural or natural, bills must be paid eventually. In nature's prairie, the bills are paid automatically and with amazing regularity. The wild forms have evolved methods for dispersing seed, recycling minerals, building soil, maintaining chemical diversity, promoting new varieties and even controlling weeds, e.g. through shading. The prairie has been successful because close attention has been paid to seeing that these jobs get done. Most biologists believe that natural selection alone was up to these tasks, and that purpose was not necessary.

This "no-free-lunch law" applies just as much to human agriculture as it does to the biotic cultures of nature. For when agricultural humans substitute their annual monoculture on this land, be it corn, wheat, milo sorghum, rye, oats or barley, the same bills have to be paid or failure is inevitable. Mechanical and commerical preparation of the seed and planting, the application of fertilizer, chemical and power weeding, mechanical soil preparation, pesticides and fungicides and plant breeding are all clumsy inventions we have devised for paying the same bills nature pays.

In contrast to the system of nature, which relies solely on the daily allocation of solar energy, in the industrialized world our inventions for the successful monoculture of the annuals require the stored light of the geologic past. Efficiency in energy use is the way of nature, not of industrialized people.

I mentioned at the beginning that the human-nature split was at its most dramatic in the manner in which land is covered with vegetation. Soil loss certainly ranks alongside our other major problems in importance. As a resource, soil is every bit as depletable as petroleum and the consequences of depletion are worse for more of earth's people. An over-populated planet can fare better without fossil energy than it can without soils to grow food. If soil loss were not such a reality, it would be much more difficult to argue that the way of nature is inherently better than the way of the agriculturists in the developed world. Energy use is not the major consideration.

The monoculture of annuals leads to soil erosion. The methods almost inherent in the monoculture of annuals require that ground be devoid of vegetation for too long a time, often during critical periods of the year. The forces of wind and rain can now rapidly move soil seaward. Even during the growing season, especially for the row crops, the loss is substantial. Crops such as corn, cotton and soybeans have much of their holding power destroyed between the rows as the farmer loosens the earth to cultivate. For this reason, J. Russell Smith called corn, "the killer of continents . . . and one of the worst enemies of the human future."[3]

The polyculture of perennials is another matter, however. The more elaborate root system is an excellent soil binder. It has been estimated that before Europeans came, fires were sufficiently common and any given area became burned at least once in a decade.[4] Though the top organic matter may have been absent for brief periods, the roots at least were alive and binding the soil.

What Will Nature Require of Us?

It seems doubtful that nature will uncompromisingly insist that the polyculture of perennials is the only way humans can peacefully co-exist with her. As I mentioned earlier, she employs some annuals in her own strategy. One might begin a limited systematic inquiry into the nature of a high-yielding sustainable agriculture by asking whether it is the annual versus perennial condition of the plant or monoculture versus polyculture we need to investigate first.

In a more thorough-going systematic study, we may have to contrast, not just annual versus perennial, or monoculture versus polyculture, but the woody versus the herbaceous condition and whether the human interest is in the fruit/seed product or the vegetative part of the plant. When we consider these four contrasting considerations, in all possible combinations, we have sixteen categories for assessment.

We can eliminate four of these sixteen categories listed in the Table for they involve woody annuals, an unknown phenomenon in nature. This leaves us with twelve categories for consideration. Eleven of these remaining combinations are currently employed in the human enterprise. But category seven, which involves the polyculture of the herbaceous perennials for seed/fruit production, is almost the opposite of our current high-yielding monoculture of annual cereals and legumes.

Fruit/seed material is the most important plant food humans ingest. This is so because of the readily storable, easily handled, highly nutritious nature of the seeds we call grains. Unfortunately, none of our important grains are perennial, or when they are, as with sorghum, basically a tropical plant, they are not treated as such. If a few of them had been, we might not have so thoroughly plowed from the edge of the eastern deciduous forest to the Rockies. Where we did not plow or where we did plant back nature's herbaceous perennials in polyculture, our livestock have become fat on the leaf and seed products. Throughout this entire expanse, the mixed herbaceous perennials have not been cultured for the purpose of harvesting the seed except for the occasional times when collections were made to plant more mixed pasture.

In the eastern tall grass region, the white settler substituted the domestic tall grass, corn. In the middle or mixed grass region, he substituted a domestic middle-sized grass, wheat. Part of the problem of the Dust Bowl is that we tried to substitute the middle-sized grass in what was short grass prairie.

The Dust Bowl followed the great plowing of the teens and twenties. When the dry winds blew in the thirties, the bad reputation for the region became firmly implanted on the American mind. We have had other severe droughts in the area since, and the wind has blown just as strong. All the work done by the Soil Conservation Service and others to prevent other dust bowl conditions should be applauded. It is truly the work of thousands of diligent and dedicated people who have spent most of their productive lives thinking and working on the problem, but still the most sobering fact cannot be ignored.

Poly vs. Monoculture	Woody vs. Herbaceous	Annual vs. Perennial	Fruit/Seed vs. Vegetative	Current Status
1. Polyculture	Woody	Perennial	Fruit/Seed	Mixed Orchard (both nut & fleshy fruits)
2. Polyculture	Woody	Perennial	Vegetative	Mixed Woodlot
3. Polyculture	Herbaceous	Annual	Fruit/Seed	
4. Polyculture	Herbaceous	Annual	Vegetative	Dump Heap Garden,* Companion Planting
5. Polyculture	Herbaceous	Perennial	Fruit/Seed	
6. Polyculture	Herbaceous	Perennial	Vegetative	Pasture & hay (Native or Domestic)
7. Monoculture	Woody	Perennial	Fruit/Seed	Orchard (both nut & fleshy fruits)
8. Monoculture	Woody	Perennial	Vegetative	Managed Forest or Woodlot
9. Monoculture	Herbaceous	Annual	Fruit/Seed	High-Producing Agriculture (wheat, corn, rice)
10. Monoculture	Herbaceous	Annual	Vegetative	Ensilage for Livestock
11. Monoculture	Herbaceous	Perennial	Fruit/Seed	Seed Crops for Category 12
12. Monoculture	Herbaceous	Perennial	Vegetative	Hay Crops (Legumes & Grasses) & grazing

*See *Plants, Man & Life* by Edgar Anderson for the splendid chapter on Dump Heap Agriculture

The soil is going fast. On some flat land there may be very little loss, but on rolling land the loss can be as high as sixty tons per acre per year. According to the General Accounting Office, the average yearly loss is 15–16 tons per acre. Based on a random sample, eighty-four percent of the farms are losing more than five tons of soil per acre each year.[5] Furthermore, there is little difference between farms participating in USDA programs and those which do not.

Unless the pattern of agriculture is changed, our cities of this region will stand as mute as those near the Great Wall of China, along the fertile crescent or the northern region of Egypt which once hosted grain fields that supplied the empire of ancient Rome.

If we are serious in our intention to negotiate with nature while there is still time for Americans to heal the split, are we not being forced to ask if nature will uncompromisingly require us to put vegetation back on the ground with a promise that we are not to plow except for the occasional replanting? If that is nature's answer from the corn belt to the Rockies, will it require that we develop an agriculture based on the polyculture of herbaceous perennials which will yield us seeds not too unlike our cereals or legumes? This category, so glaringly blank in our table, needs filling desperately; and yet to contemplate the research, breeding, establishment of the crops, the harvest and separation of seeds is mind boggling. All this effort must go hand in hand with the transportation, milling and ultimately, the eating of this "instant granola in the field."

Is it too much to expect plant scientists to come up with such perennials, either through some inter-generic crossing of our high-producing annuals with some perennial relatives, or by selecting some wild perennial relatives which show promise of a high yield of a product that is at once abundant and tasty? Any scenario surrounding such an agriculture does seem to truly belong to a fantasy world. For mechanized agriculture it would mean either a minimum amount or a complete absence of plowing, disking, chiseling and mechanical power weeding. There would be only harvest, some fertilizing, pest control, genetic selection and the occasional replanting.

How about fruit or nut trees for saving our soils? The virtues of orchards need little promotion. Trees have many advantages over their herbaceous relatives. Propagation by twig or bud through grafting allows the multiplication of any useful mutant by the millions. Because of their deep rootedness and woody nature, trees are better able to withstand fluctuations between drought and rainfall. Trees provide shade and fuel and are altogether handsome on the landscape. Because of their numerous attributes, I would encourage any lover of trees to commit all the acres he or she could to their culture.

In spite of all the gifts trees offer, the fact remains that year in and year out they do not compete with wheat, rice, corn and soybeans either in ease of production and harvest or yield. Even if we discount how long it takes for trees to bear fruit, any compelling substitute must lend itself to machine harvest and high production. Cultural tradition may be partly responsible here but even so there are few substitutes for grains.

Trees can compete in saving soil, however. There are a wide range of slopes, from the extreme of steep, rocky land all the way to the flat alluvial, with soils which can best be protected by orchard trees, if not returned to the wild. A land ethic stronger than a short-run production ethic will favor trees over high-yield annuals on sloping ground.

How feasible is the development of high-yielding, seed-producing herbaceous perennials for this century? The answer to this question hangs on settling a fundamental question of plant science, the question of whether perennialism and high-seed yield are mutually exclusive or not. Some highly reputable plant geneticists I have asked, who have worked and thought on the question, not only have discouraging comments but lean toward a categorical "no" when asked about the possibility of co-existence of perennialism and high yield. Others felt there were some possibilities. We began to explore this question at The Land Institute by going to the literature and later conducted direct investigations of our own. Smith reports that the Millwood variety of the honey locust tree will produce 1800 pounds of fruit/seed material on a per acre basis.[3] At The Land, smooth sumac, a shrub, with no human-directed selection will produce over 1000 pounds of fruiting material per acre. Another shrub, sand plum from Dickinson County, Kansas, on a per acre basis, as determined by extrapolation from a 10.5-square-foot area, will yield 6480 pounds of fruit/seed material. When the pit is removed, the dried flesh will yield 3820 pounds! We can now be satisfied that perennialism and high yield are not mutually exclusive.

The next question is whether *herbaceous* perennialism and high yield are mutually exclusive. At The Land we have grown several such species in 5-meter-long rows and have extrapolated to a per acre basis, recognizing that such a method is likely to yield overly optimistic results. A three-year-old plot of *Ratibida pinnata*, a perennial member of the sunflower family, yielded 1470 pounds of seed. This plant is important for it contains the two important fatty acids, linoleic acid and linolenic acid, which occur in oils and are essential in animal nutrition. Maximillian sunflower in the first year yielded 1300 pounds but only 400 pounds on the 3-year plot. Nevertheless, these results from wild species in which little, if any, selection has taken place are encouraging. A critic might point out that these are all members of the sunflower family and that we cannot be considered serious until we look to the grasses and herbaceous legumes. But this is jumping ahead a bit for we have sought to answer several questions about perennialism and high-yield being mutually exclusive, one at a time.

As we have just seen, it required but minimum effort to show that they were not mutually exclusive when we consider woody species. Therefore, it is not a condition of *flowering plants*. Furthermore, our results with a few herbaceous members of the compositae or sunflower family suggest it is not mutually exclusive when we consider the *herbaceous* condition. At this point it is important to establish some minimum standard for what constitutes a high-yielding crop. Since 30 bushels per acre has long been regarded as a good wheat crop, at 60 pounds per bushel, an 1800-pound yield for any herbaceous perennial would be a high-yielding crop. As we have seen, at least a few members of the sunflower family approach this. Certainly grasses and legumes must have priority in new crop development and we are fortunate to find a few encouraging results in the literature. Perhaps most noteworthy is the work of Dr. Robert M.

Ahring at Oklahoma State University. He has managed to produce yields in buffalograss of 1727 pounds/acre/year under certain fertilizer treatments.[6] Over a four year period his plots have averaged 1304 pounds of seed burs per acre. Of course, much of the bur could not be considered edible. Nevertheless, it is fruit/seed material. The grass *Sporobolus cryptandrus*, with no selection, is reported to have yielded 900 pounds an acre.[7] Essentially all 900 pounds are edible.

Until now, most of the work on new crop development has involved the crossing of economically important annuals with some of their wild perennial relatives. As mentioned earlier, there is a tendency for older patches to decline in productivity. It is well known that perennials provide a fair to good stand the first year and often a pretty good stand the second, but by the third year, production is headed steeply down. Though this is certainly a problem, we can't say for sure that this will always be the case and that this need be a biological law any more than an average of 40-bushel-per-acre corn in the pre-hybrid days was immutable biological law. It may have been a biological reality, but plant breeders proved it was not a biological law.

There are numerous problems associated with the development of an agriculture based on perennials. Nevertheless, such a program would seem to have more promise than the effort to convert temperate grasses into legume-like, nitrogen-fixing cereals. At least the high-yield cereals we tested have numerous perennial close relatives, while the base line for any encouragement in the development of nitrogen-fixing cereals are the few wild tropical grasses in which minimum amounts of nitrogen are fixed.[8]

In spite of all the discouraging events certain to confront the plant science community, the development of only one new high-producing perennial crop could pay dividends for both developed and developing nations forever. Such a crop could go a long way toward preventing erosion and desertification, problems common to both types of nations.

The important point for our consideration here is that no new breakthroughs are necessary for us to begin a very large program now, involving scores, if not hundreds, of crosses and selection experiments in our universities and research organizations. Some incentive seed money is always needed to accompany policy change. But we need not wait for additional scientific and technological developments. These developments occurred earlier in our century as biologists sought to fuse Charles Darwin's ideas of evolution through natural selection with Gregor Mendel's principles of heredity, both of which had developed over thirty-five years before. This exciting period of history in biology, whose excitement we too readily forget with our contemporary mania over such gee-whiz genetics as cloning and genetic surgery, began early in the century with attempts to establish the chromosome theory of heredity and by and large culminated with the elucidation of the chemical structure of the hereditary material, the DNA, by Watson and Crick. During this period, techniques were developed to count chromosomes and follow them through the various stages of replication and division. Chromosomes were irradiated, broken and fused and some of their genes mapped. Sterility barriers between species came to be understood, and artificial hybrids, including some resulting from intergeneric crosses, were successfully made. We came to understand how species arose through chromosome numbers being doubled or reduced, and investigators learned to artificially induce these changes. Chromosome numbers have been successfully doubled through chemical agents to the point it has become a matter of routine. Numerous species have had their karyotypes or genetic fingerprints determined.

The work that linked the independent ideas of Darwin and Mendel is now a reservoir of practical knowledge. Interestingly enough, a relatively small amount of this information has been used in crop and livestock improvement. The plant scientist and the breeder did use some, but mostly they applied the tools from the newly-emerging field of statistics and made significant advances in crop production through improvement in experimental design and a better understanding of hybrid vigor. These assiduous experts were less interested in new crops than in their imaginative programs of "fine-tuning" the traditional crops. Thankfully, the same hardware (optical equipment, growth chambers, greenhouses etc.) and the basic research necessary for "fine-tuning" will be needed as we research the fundamental question of whether herbaceous perennialism and high yield are mutually exclusive or not. The supporting fields of plant physiology, plant pathology, entomology and biochemistry have the necessary working bibliographies, equipment and experts to work in concert with the geneticists to gain information which has an impact on our national problem—soil loss.

This is a period in which we should encourage much wide-ranging imagination and speculation on new crop development. Numerous botanists and crop scientists will have plant candidates in mind. I have already mentioned several high-yielding wild perennial species, but only to shed light on the biological question as to whether herbaceous perennialism and high-yield are mutually exclusive or not. Species that currently have lower yields, may be more important for development in the long run. At The Land Institute we are working on Eastern Gama Grass and the newly discovered perennial corn from Mexico. We have chosen Eastern Gama Grass, *Tripsacum dactyloides*, because it is a perennial relative of corn and a plant that cattle relish. Its seed consists of 27% protein and it is nearly twice as high in the amino acid methionine as corn. This high protein percentage, three times higher than corn and twice as high as wheat, should allow the breeder to sacrifice considerable protein content in the push for higher yield. Right now the yield is around one bushel per acre, but there are several characteristics, which if put together might increase yield many fold right off. (1) It has already been extensively studied, particularly by those interested in the evolution of corn.[9] Therefore, important basic information already exists. (2) The species is already at home in our corn belt for it nearly rivals corn in the extent of its distribution, ranging from Florida to Texas and Mexico north to Massachusetts, New York, Michigan, Illinois, Iowa and Nebraska.[10] (3) Because this tall, stout perennial has thick rhizomes, any desirable races could be propagated vegetatively from clumps. (4) The part of the flower that sets seed is localized and separate from the part that produces pollen. Therefore, no tedious effort is necessary for the breeder to emasculate before making crosses. (5) The species contains two more or less true breeding chromosome races. The virtues of this species are indeed numerous.

Perhaps just as promising as the above mentioned corn relative are some of the relatives of our high-yielding, leguminous, nitrogen-fixing crop, soybean. *Glycine max* (L.) Merr., as it is scientifically called, is an annual. Most species in this genus are perennial. The genus itself consists of three subgenera which include 10 species and 18 genetic entities, i.e. subspecies or varieties. Furthermore, there are three closely related genera comprising some 12 additional species.[11] The variation within *Glycine* alone is truly remarkable. However, the entire American soybean industry, which produces 75% of the world's supply, in the words of Professor Jack R. Harlan of the University of Illinois, "can be traced to six accessions introduced from the same part of Asia."[12] It would seem that something could be done to test our basic question concerning perennialism and high-yield with some of the other species of this genus or even the relatives of the closely related genera.

The third example involves a grass again, the Panicum complex, which includes broomcorn millet or Hog Millet, as it is sometimes called. Most species are perennials and the genus *Panicum* has a large range both in latitude and longitude, suggesting great genetic elasticity. A closely related genus *Setaria* includes the Common Millet as one of its species.

These are but three examples. The possibilities are there for other groups as well. All that is needed now is interest on the part of investigators and some seed money from foundations and the government for researchers to redirect their efforts.

Why have we not developed any new herbaceous perennial seed crops so far? One explanation might be that the development of a suitable crop is beyond us.

Another might be that we have lacked, in the right places, the kind of holistic thinking that would link the high-yield seed production of annuals with soil loss. Even if we have seen the problem, most of us must confine our breadwinning efforts to a narrowly-defined job description. In discussing the problem with colleagues, I have found that many, like myself, were aware that soil was probably being lost at an unacceptable rate, but were not aware until the release of the General Accounting Office's study that the problem was so acute and therefore we had not concentrated our minds on the need for an agricultural solution.

New crop development has had relatively little attention in the history of our species since eight to ten thousand years ago when several generations of the most important revolutionaries ever to live on earth gave us essentially all of our crops and livestock. Of the thousands of seed-producing plant species known, fewer than one percent have been utilized by humans for food, clothing and shelter. By and large humans do the easy things first, and so our crop scientists have improved the plants that have already demonstrated their amenability to cultivation.

Of the hundreds of crops available in our inventory now, fewer than a dozen supply the huge bulk of food stuffs. Because these plants have an economic history, there is a ready-made economic data base for evaluating market opportunities against cost for any breeding work to be done. After all, much of our culture is built around relationships involving the farmer, the processor and the consumer. There has always been plenty of work to do in crop improvement without looking for more.

Therefore, we have logically questioned the wisdom of adding more plants when we are not fully utilizing many of the proven plants which are already available.

There is probably another reason why we have not looked to herbaceous seed producers to save our soils and yield high-quality food. Imagine the psychological climate of the scientific community forty years ago. We were still in a depression, and the dust storms had already become legend. We have seen what a dramatic response the Roosevelt administration made to this problem. The high caliber people Bennett employed, the good reputation quickly gained, all of this allowed scholars and laymen alike to turn their attention to other matters, entirely confident that the effort to save the soil was in the best hands possible.[13] It would simply be a matter of time before this problem was solved. Since the procedures were both practical and scientific, everyone felt comfortable. There was little incentive to look elsewhere for solutions to the soil loss problem.

What environmental benefits (other than reduced soil loss) could we expect from an agriculture of herbaceous perennial seed-producers?

Perennial culture could reduce energy consumption. The energy for traction in seed bed preparation and cultivation is significant; it comprises the major fuel bill for the farmer year in and year out.

Perennial culture could reduce pesticide dependency resulting in both energy savings and healthier soil and food. As mentioned earlier, the direct fossil fuel energy that goes into our pesticide program nation-wide is at least eighty percent of the one billion pounds sprayed on our fields each year.[14] This amounts to around two million barrels of oil. (Not included in these figures is the energy cost for making the chemicals, nor distribution to the farmer, nor his energy cost for application.) Because many of these new crops would presumably be the result of inter-specific and inter-generic crosses, they could represent a broad genetic base of disease resistance. The current "hard agricultural path" promotes a genetic narrowing and therefore increased vulnerability to pests overall.

Perennial culture would reduce our dependency on commercial fertilizer. I assume this because the application of fertilizer to perennial forage crops is, on the average, much less than to annual grain crops. The slow decay of plant materials from perennials releases nutrients at a rate that new growth can more efficiently assimilate. This saving would be significant, for not only is commercial nitrogen fertilizer energy-intensive, as we have seen, it is toxic to children and farm animals.[15] It is not uncommon for water tables to have high levels of nitrates and for aquatic ecosystems to be placed greatly out of balance. Besides, a real fertilizer crisis could develop. The feed stock for much of our commercial fertilizer is natural gas, and in 1974, twenty-two percent of the interruptable supply of natural gas was devoted to the manufacture of fertilizer.[16]

The development of new, high-yield perennial, seed-producing crops could reverse the current decline of our domestic genetic reservoir. Population increase and intensive agriculture have reduced the amount of "waste land" where teosinte, the wild relative of corn, once lived. For wheat and rice, too many of the old low-yielding but faithful varieties of various races and ethnic groups have been driven from the fields.[17] Many of these are low performers by modern standards, but have been the genetic bank which breeders would tap now and then to introduce new germplasm into crops made narrow by selection.

The cost for maintaining a very wide spectrum of genetic variation is prohibitive for most of the seed companies. The National Seed Storage Laboratory at Fort Collins, Colorado is charged with the expensive, and difficult responsibility of keeping genes stored. The most efficient storage is in living organisms.

In summary, success in herbaceous perennial crop development would lead to a reduction in resource depletion for both fossil fuels and germplasm and would reduce pollution of our waters, soils and ultimately ourselves. Even if we are not successful in our attempts to develop high-yielding herbaceous perennial crops the low-yielding and otherwise useless new stocks may serve as a bridge for introducing new germplasm into our high-yielding annuals.

Perennial Polyculture

The development of perennial seed producers to rival our annuals is not the most radical kind of agricultural research needed in the affluent countries. Traditional cropping systems have widely employed polycultures already. Eventually, the polyculture of perennials to meet a variety of human needs in the developed world will be necessary. Such a program may require 50 years or more before we see combines going through our fields harvesting seeds of three or more species, including legumes. These plants will have been selected to mature in synchrony and will accommodate themselves to machinery yet to be designed by agricultural engineers.

I have already discussed whether perennialism and high yield are mutually exclusive. A second basic consideration has to do with the comparative yield of a perennial polyculture compared to a perennial monoculture in temperate regions of the globe. Numerous ensembles will have to be matched according to flowering and seed set time, to be taken into consideration along with root type and depth of penetration.

Our third and final major category of interest for the moment is related to the first two questions. We want to determine if several families of plants which are common to the native prairie could make yield-increasing contributions to our polycultures. For example, we know that legumes fix at least a minimal amount of nitrogen on the prairie, but what is the role of the sunflower family, or the rose family, the mint family, etc.? Perhaps they play no role in the overall health of the prairie ecosystem and are just there. In fact, even the value of the legume family might be brought into question because most of the nitrogen available to the prairie falls with the rain and snow.[18] Perhaps legumes are allowed not because of nitrogen fixation abilities which ultimately contribute to the entire plant community, but because they have niches somewhat independent of the other species.

Perennial Polyculture and Crumb Structure

Continued soil fertility depends, to a large extent, on plant associations which form and preserve an appropriate crumb structure.[19] A crumb is a granular substance resulting from various chemical and physical factors at work in a field and is of prime importance in all highly productive land ecosystems. Without these crumbs some soils become powder, while others consisting of clay change to a butter-stickiness when wet.

Crumb structure declines in year-in-year-out monoculture or where there is too much cultivation, too much fertilizer and too little humus. Knowledge of the conditions that promote this formation lends further support to the idea of perennial polyculture, in particular the perennial grasses. Associated with the grass roots are dead rhizosphere bacteria, which act as binding agents. But grasses have other virtues to assist crumb presence in the soil. Crumbs are less exposed in a grass field when it rains and grass roots exercise a wick effect, pulling water to lower levels. This downward allocation of water keeps the crumbs in the upper levels from slurrying off. A grass mat would also protect crumbs from the dessicating effects of frost and sunshine. The roots of annual grasses doubtlessly promote crumb formation, but strong roots, more likely to be found in perennials, produce more crumbs than grasses with smaller, more fibrous roots.[20]

Even when we are considering something as lowly as crumb formation, it seems inappropriate to simply swap annual monoculture for perennial monoculture. One reason is that clover in monocultures is less effective than grass in monoculture in producing crumbs. But when clover and grass are grown together the crumb-forming action of the grass increases.

After crumb structure has been drastically reduced in a field, redemption is not immediate. From four to thirteen years, depending on soils, are required before increase in crumbs can be detected.[20]

Why these humble, irregularly sized and shaped crumbs are such a strong factor in the soil environment begins to make sense when we contemplate how they affect soil texture. They create porosity, opening the soil so effectively that abundant water readily drains into the sub-soil. This granular soil allows the nutrients, oxygen, and carbon dioxide to readily diffuse in and around roots. Crumbs make it mechanically easy for roots to grow and worms to burrow. Crust formation or glazing is prevented and when dry weather does come the small soil particles are sufficiently held together to minimize blowing.[20]

In a successful herbaceous perennial polyculture, we would expect that soil erosion would cease to be a problem. Because of the chemical diversity of such an ecosystem, insects and plant diseases would also be less serious. In a natural prairie polyculture, weeds are managed by the shading system. Nutrient balance is also managed by the system with little human involvement. Water is held by the spongy mass and a deeply penetrating root system has a wick effect and "pulls" water down when the heavy rains occur. Once a balanced polyculture has been planted, soil preparation and preparation of seed and planting are all done by the system.

This is not to say that numerous problems do not await such an agriculture. Long before we get to the point of fine-tuning such a system our farmers will need to employ the entire array of sound soil conservation measures as a holding action.

It has always been hard to see where we stand in history as we are living it. Could it be that we have promoted monocultures of grains when possible because agriculture is back-breaking work? There have been advantages to concentrating our food energy as much as possible. Perhaps we can now think seriously of machine harvest of polycultures of grain on a massive scale. The paradox is that we can now live in ecosystems more like the gathering-hunting ones in which we evolved because of high technology. We can almost "go home again," finally, because we have the technology which will allow us to live there.

A Possible Cultural Barrier—Food is such an intricate part of cultures that we should not expect rapid adoption of many of these new crops into our diet. Humans have historically been rather conservative about food habits. There is one characteristic, however, that describes the human animal. We are grass-seed eaters and, secondarily, legume-seed eaters. I would expect that any shift as far as texture and taste is concerned, would be no more drastic than the shift from rice to wheat or from wheat to rice. Nevertheless, there will be shifts that people may be reluctant to make. A major concern in all this is the importance of the sticky, nitrogenous substance, gluten, which allows bread dough to be stretched without breaking apart before making a beautiful loaf.

Different wheats have different flavors and other characteristics that determine the products that are made with them. The entire array of such characteristics which have already been adopted by humans may, in fact, capture much of the range of variation among species from wide crosses yet to be made. I see no reason why we should not approach the whole program with a sense of adventure—especially if we recognize the ecological benefits of growing the crops.

Overall, I am optimistic that there will be no meaningful cultural barrier to their adoption. If only a hundredth of the advertising is applied to the promotion of eating these healthful grains that is applied to the array of unwholesome junk food we ingest now, no cultural barrier can stand in the way of their whole-hearted adoption.

How Does the Proposal Discussed Here Relate to the Larger Vision of the Environmentalist?

Because sunshine is dispersed rather evenly over the earth; because nature's three dimensional solar collectors, called green plants, are also dispersed; because these collectors are so critical to all other life forms, including humans; because the land for growing these collectors in the U. S. is eroding at the rate of nine tons per acre per year on the average: any who advocate a sunshine future or soft energy path must ultimately adopt a land ethic which embraces an energy ethic.

The soft energy path or sunshine future advocated by Amory Lovins, it would appear, would ultimately require a decentralized society. Sunshine is dispersed. Nature's three dimensional solar collectors are dispersed. A major emphasis of Lovins' thesis is the thermodynamic match, i.e. energy source and energy end-use should be matched. Therefore, should not nature's people be dispersed? It seems reasonable enough—but do we have enough information to say with any high degree of confidence, what the distribution pattern should be? First off, *how* people live, in terms of consumption, is more important than *where* they live, in the first round of growing scarcity.

Let us assume for the moment that our systems analysts widely agree that eventual decentralization is absolutely necessary for a sustainable life on the planet. To build the infrastructure for decentralization may be both energy and materials intensive. The romantic back-to-the-land movement, as minimal as it is thus far, is a signal of something desirable. But what happens when we all get there, after the first generation of back-to-the-land romantics have been buried organically in their gardens? Will their children maintain the back-breaking work most humans have sought to avoid over the centuries? Isn't this one of the components of the human condition? It has yet to sink into our culture that we are still basically gatherers and hunters, and that the era of agriculture is but a thin veneer over an evolutionary past which tolerated a great deal of leisure. The appeal of the countryside is the appeal open space has always had for us gatherer-hunters. The appeal of the city is that it at least faintly suggests a mixture of leisure and stimulation most of us need. R. B. Lee has reported that only 65% of 31 !Kung bushmen in Botswana spent two to three days each week gathering food while "35% of the people did no work at all."[20]

Van Rensselaer Potter has pointed out that we all have a need for an optimum stressor level which varies for each of us.[21] On one side of this optimum is boredom, which can come from too long a period in the fields. At the other end is the problem of information overload which may come from being over-stimulated in the city. The only way I can see the decentralized culture joyfully surviving is for our technology to allow us both stimulation and time for leisure, so that we might play out the longings of that gathering-hunting body and brain.

But what do perennial seed-producers have to do with a Utopian vision—or with leisure and stimulation? In the Summer, 1949, issue of the now defunct *Land Quarterly* is an article entitled "Sweet Living at Yellow River" written by a Channing Cope. Mr. Cope describes a goal which he and his family had recently achieved on a farm in Georgia:

> At long last we have it. The result is far beyond our fondest dreams, for we never thought we would so utterly eliminate drudgery from farming

> through a combination of four basic plants working in natural unison to sustain life. These four plants are Kudzu, sericae lespedeza, Kentucky 31 fescue grass and ladino clover. . . Front porch farming calls for perennials to the greatest extent possible, and if these are not possible we court the annuals which have the habit of reseeding each year. Therefore, no plowing except to get the crop started.

Speaking of anyone who would have such a farm, he continued that such a person

> will pass it on to the next generation in better shape that he found it. He couldn't design a better monument than a weather-proof farm. It won't wash away. It won't wear out. It will furnish basic food. It gets better as the years move on. It makes a local, statewide and national contribution, and, to the extent of its influence, it helps prevent war.

Channing Cope's perennials, of course, are all plants devoted to the production of vegetative material. His personal dream could come true because he lived in a culture where the slack was taken up by the thousands of seed-producing farmers of our nation. Nevertheless, what he calls "front porch" farming as an appealing way of life brings to mind the need of the gatherer-hunter. It also illustrates that such leisure has provided Channing the opportunity to reflect on the old religious questions and critical values necessary for a sustainable culture and agriculture when he speaks of building a "weather-proof farm" as his "monument." He saw in his individual action a positive chain of events, perhaps as part of a web, but certainly as a local response to large problems which even included the prevention of war!

The scientific-technological revolution has surely already provided us with enough recyclable hardware to keep a decentralized society stimulated and in touch with one another at home through the telephone, television and perhaps even the home-based computer erminal.

But the human-nature split remains. As immodest as it may sound, I think we can at once provide leisure and begin to close the split if the Channing Copes can be provided with high-yield, seed-producing herbaceous perennials in polyculture.

I have mentioned that the chemotherapy treatments to the land promote a temporary vigor more impressive than our fields have ever known. Though the physician may rejoice with his cancer patient that he is feeling better in response to the treatment, he is also careful to monitor the telltale systems of the body. Similarly, those interested in the long-term health of the land need only stand on the edge of a stream after a rain and watch a plasma boil and turn in the powerful current below and then realize that the vigorous production of our fields is, unfortunately, temporary. Since we initiated the split with nature some 10,000 years ago by embracing enterprise in food production, we have yet to develop an agriculture as sustainable as the nature we destroy.

References and Notes

1. From the novel *Cry, The Beloved Country*, 1948. Charles Scribner's Sons, New York.
2. This first century writing is in Lucium Janius Moderatus Columella, "on Agriculture," with English translation by Harrison Boyd Ash, Cambridge, Mass., 1942.
3. J. Russell Smith, 1953. *Tree Crops*. The Devin-Adair Company, New York.
4. Dr. Lloyd Hulbert, Plant Ecologist at the Kansas Agriculture Experiment Station and Professor of Biology at Kansas State University, after observing the time in which woody vegetation encroaches when fire is not present, has supplied me with this number. In the eastern part of the prairie, it would be more frequent than ten years. In the western third of the grasslands, 15 years or more could elapse without fire. Grasses have probably evolved to invite fire.
5. In February, 1977, the General Accounting Office (GAO) released an analysis of the United States Department of Agriculture (USDA) soil conservation efforts. The GAO based its conclusions, in part, on visits to 283 farms in the corn belt, Great Plains, and the Pacific Northwest.
6. Robert M. Ahring, May, 1964. "The Management of Buffalograss for Seed Production in Oklahoma," Technical Bulletin T-109. Oklahoma State Experiment Station.
7. H. Ray Brown, 1943. "Growth and Seed Yields of Native Prairie Plants in Various Habitats of the Mixed-Prairie," Transactions Kansas Academy of Science, Vol. 46, pp. 87–99.
8. Döbereiner, Joanna, 1977. "N Fixation Associated with Non-Leguminous Plants" in *Genetic Engineering for Nitrogen Fixation* edited by Alexander Hollaender. Plenum Press. p. 451.
9. See, for example, the paper by J.M.J. de Wet and J.R. Harlan delivered at the Symposium on Origin of Cultivated Plants at the XIII International Congress of Genetics. There are numerous literature citations which give one a sense of the history of studies on *Tripsacum*.

10. Julian A. Steyermark, 1963. *Flora of Missouri,* Iowa State University Press, Ames, Iowa. pp. 252–254.

11. J.M. Herman, December, 1962. "A Revision of the Genus *Glycine* and its Immediate Allies." U.S.D.A. Technical Bulletin No. 1268.

12. Jack R. Harlan, 1972. "Genetics of Disaster." *Journal of Environmental Quality,* Vol. 1, no. 3.

13. Wellington Brink, 1951. "Big Hugh's New Science," *The Land,* Vol. X, no. 3.

14. Steven D. Jellinek, December, 1977. "Integrated Pest Management from Concept to Reality." U.S. Env. Protection Agency.

15. Paul R. Ehrlich, Anne H. Ehrlich and John P. Holdren, 1977. *Ecoscience: Population, Resources, Environment*. W.H. Freeman & Co., San Francisco. p. 558.

16. Guy H. Miles, November 1974. "The Federal Role in Increasing the Productivity of the U.S. Food System." NSF—RA-N-74-271, p. 20.

17. Garrison Wilkes, 1977. "The World's Crop Plant Germplasm—An Endangered Resource." *Bulletin of the Atomic Scientists*. See also "Our Vanishing Genetic Resources" by J.R. Harlan, reprinted in *Food: Politics, Economics, Nutrition and Research*. AAAS edited by P.H. Abelson, 1975.

18. Robert G. Woodmansee, July, 1978. "Additions and Losses of Nitrogen in Grassland Ecosystems" *BioScience*. Vol. 28, No. 7, pp. 448-453.

19. N. Pilpel, May, 1971. "Crumb Formation," *Endeavour* 30:110.

20. R.B. Lee. 1969. "!Kung Bushman Subsistence: An input-output analysis." *Bull. Natl. Can*. 230:73-94.

21. Potter, Van Rensselaer, 1971. *Bioethics: Bridge to the Future*. Prentice-Hall, New Jersey.

11
Outside the Solar Village: One Utopian Farm

> Many of the most lively, intimate expressions of spirit spring from the joyous, continuous contact of human beings with a particular locality. They feel the age-long spirit of this valley or that hill each with its trees and rocks and special tricks of weather, as the seasons unfold in their endless charm. If life can be made secure in each community and if the rewards of the different communities are distributed justly, there will flower in every community not only those who attain joy in daily, productive work well done; but also those who paint and sing and tell stories with the flavor peculiar to their own valley, well-loved hill, or broad prairie . . . Every community can become something distinctly precious in its own right. Children will not try to escape as they grow up. They will look ahead to the possibility of enriching the traditions of their ancestors.
>
> —Henry Wallace[1]

The year is 2030 in a world with increased consciousness. People everywhere—on farms, in villages, and in cities—have sustainability as their central paradigm. They think globally and act locally. Regional semi-self-sufficiency is emphasized but the principles of the New Age Farmers are the same from New England to Southern California.

Our utopian farm is in Kansas, below the 39th parallel and east of the 98th meridian. The area averages about 28 inches of rainfall each year but the evaporation is in excess of rainfall. This is farming country which, before being plowed more than a century ago, was a biotically rich landscape. Stories handed down through the grandparents tell school-age children how the breaking of this virgin sod sounded like the opening of a zipper. A few miles east is the western edge of the vast Tallgrass Prairie, dominated by such species as Big Bluestem, Indiangrass and Switchgrass. Scarcely thirty miles to the west are the mixed prairies dominated by Bluestem and Sideoats Grama.

Because of the minimal landscape relief, the Great Plains is one of the few regions where it makes sense to divide the land into one mile square parcels. A road surrounds almost every square mile. This is a land which, after the "Great Plowing" in the early 1900's, supported such high-producing annual crops as wheat, sorghum, milo and soybeans. Before 1990 only native pastureland and roadsides carried the grasses that were characteristic of the region before the European arrived. This prairie land, mostly because of forced grazing, has long since lost 20 or 25 native prairie species. What was left was not prairie but grassland. During most of the last century, wheat was an important export crop for the region; we are fortunate that even more grassland wasn't plowed. Church leaders, farmers and grain men had said that we must sell grain to feed a hungry world. This was a moral veneer over a basically economic consideration, but it was enough to discourage the initial development of mixed perennials. Traditional crops were proven producers regardless of their tremendous toll on finite energy resources, finite soil and, for western corn growers, finite fossil ground water.

But now in 2030 the settlement pattern differs drastically from what it was in 1980. In this immediate area, each family lives on 160 acres, or four families per square mile (640 acres). The dwellings of most families are near the middle of the square mile section but on their own property. Therefore, within 300 yards of each other are usually 16—20 people. Their small village and main trading center, which includes both school and churches, is two and a half miles away from our farm. No one in the rural service area of the village is ever any farther away. The village's service area covers 16 square miles (four miles on each side) and includes 64 farm families totaling about 250 people.

Westward, in the mixed prairie, one half section (320 acres) is needed to support a single family, and nearly 200 miles west of the mixed grass country, in the short grass prairie, two square miles is usually necessary to support a family. Eastward and in some of the West, it is another story. Along the Missouri in Nebraska, the southwestern half of Minnesota, most of Iowa, in southeastern Wisconsin, northern Indiana, northwestern Ohio, east central Michigan; along the Mississippi in western Tennessee and northwestern Mississippi; in much of the Sacramento Valley, as well as in numerous other localized areas throughout the country, fewer than ten acres—but never more than twenty—are enough to support a family. It is not that production is always higher than in our area, it is just that a combination of factors, including rainfall, makes a sustainable yield more assured. The carrying capacity of the land is so varied that when we say the average farm, nationwide, is forty acres, we must immediately realize the limited meaning of that statistic.

Regardless of farm size, the village population seldom exceeds the farm population by more than a factor of two. An entire community in our region comprises around 750 people, including the 260 people on farms. Let us compare this to the distribution pattern nationwide. The population of the United States is around 300 million and is scheduled to stabilize completely in the next seven years, in spite of the fact that zero population growth procedures have been in effect since the early 1980's. The momentum of that past is still with us, though insignificantly so. But it is the distribution of the population which has been radically altered over the last 50 years.

Most of the major cities have experienced drastic declines and the number of cities of 40,000 or less has greatly increased. Optimum city size was widely discussed in the last century. Many of the New Age pioneers concluded, though there was nothing like unanimous agreement, that much of the social pathology of our former urban areas could be attributed to the spiritual dangers which arise when people no longer know or feel their rootedness in the land. When heat comes from a stove, food from a grocery store, building materials from the lumber yard and the automobile from the showroom floor, the spiritual loss is devastating to the society. It doesn't necessarily take a city of a million, many concluded, to provide the "critical mass" necessary to help a large number of humans live up to a broad spectrum of their innate potentialities. A population of 40,000 seems to have a special energy. When Notre Dame was begun, the Paris population was 35,000. Renaissance Florence was also around 35–40,000. Regional cities now seldom exceed 40,000 and there are somewhat fewer than 4,000 such cities totaling less than 160 million people. Of course, some of the major cities still contain a few million people, but they are mostly empty, and much of their area now produces food, clothing and shelter where concrete and stone formerly dominated the environment. The civilization was a long time in learning that, by and large, the only people who really liked the big city life were merchants and intellectuals.

Of the 300 million people in the United States, some 20 million, or about one fifteenth of the population, work in the rural areas associated with rangeland and forestry. Nearly 10 million families totaling about 40 million people are living on 400 million acres of cropland. This amounts to a little over 13 percent of the total population, well over twice the percentage of 50 years ago and nearly three times the total rural population of that time. The rural villages, however, have twice as many people as the countryside they support. I mentioned earlier that the land holdings vary drastically in size. For example, in much of northeastern Illinois, a family of four can live on five acres. This puts 128 small farms or 512 people on each square mile. This is a very high density, but the productivity of the land is the determining factor. Over 8,000 people live within the 16-square-mile rural service area. Its supporting village has over 16,000 people.

Our solar village of 500 or so is necessarily different from the northeastern Illinois village of 16,000. Aside from the differing political dynamics associated with different sizes and densities, there is a commonness of purpose best reflected in the numerous bioshelters which grow what might be described as a healthful diet, though not an abundance of calories. (The fields provide most of the protein and carbohydrates for this society and it is up to the people in the villages and cities to provide vegetables and fruits and a certain amount of animal protein, mostly from fish, in the passive bioshelters, which were pioneered by the New Alchemists in the last century.) The major differences among these villages lie in how different regions' village people work with farmers to meet the expectations of the land. A pluralistic society does not preclude the possibility of holding a common allegiance.

Neither does pluralism mean that certain patterns of young and old cannot be similar everywhere. Throughout the country, older people have the option of living in the village but their presence is cherished on the farm. Nearly all have chosen to live in the village but most return to the farm daily to assist their family and neighbors in various chores. These are the people who play the most important part in the children's education.

Most communities now emphasize the value of history, and history becomes more real when adults tell personal stories which link the past to the present. The stories are about heroes, the prophets of the solar age and the pioneers in the era of decentralization and land resettlement, and villainous corporations more concerned about avoiding liability payments legally, than protecting the environment. The older people tell of a past in which nuclear power was tried, discovered to be filled with unresolvable uncertainties, and abandoned. Many of these older people lived during what is now called the "Age of the Recognition of Limits." These former doom-watching pioneers were like the children of Israel who had escaped the grasp of the Egyptians and then wandered in the wilderness for 40 years, saddled with their own slave mentality, waiting for a new generation of free minds to develop and be fit for life in the promised land. Many of the pioneers have readily admitted their earlier addiction to all the consumer products of affluence and work hard at teaching their young the true source of sustenance and health—the land. They are living reminders that this sun-powered civilization has arrived as the result of nothing less than a religious reformation.

The strong new land ethic has resulted in a different concept of land ownership. Under the Land Trust System, land is not owned by individuals in the same sense that it was 50 years ago. Nevertheless, it can be passed on from one generation to the next, and people have a strong sense of ownership. They cannot do exactly as they please with the property. They cannot willfully pollute it with toxic chemicals, sell it off for housing developments, or in any way speculate with it. Such wasteful exploitation discounts too much of the future. Activity which is potentially destructive is prohibited by a board of non-farming elders from the village; at least half of them have been farmers for 20 years or more. The board includes two members from the regional city, and both sexes are equally represented.

On our farm, the well-insulated house is partially underground and equipped with both passive and active solar installations for hot water and space heating. Though it is 100% solar, a wood stove is in place as a back up. A water-pumping windmill and two wind-electric systems provide power for the farmstead.

A combination of technologies from the past are appropriate for the farm's water system. A water-pumping windmill brings water to tanks for the livestock and household use. Trenching machines and plastic pipe are used to deliver the water wherever needed for human convenience. One wind generator takes care of all refrigeration needs and simply cools the deepfreeze and refrigerator when the wind is blowing. Since the refrigerator itself is the "accumulator," no batteries are needed. The other wind-electric system consists of an induction motor which kicks in when the output of the wind-powered generator is greater than the load on the service line. The induction motor, which is similar to that found on washing machines in the 1930's, is plugged into the wall receptacle and runs the kilowatt-hour meter backward, giving the farmstead an electrical energy credit. A special meter records the number of hours generated. If this household wishes to break even on the utility bill, its unit must provide four kilowatts of electricity to a privately-owned utility for each one it receives. There is just enough electricity generated in the area from both wind and low-head hydroelectric turbines to supply the needs of the countryside, village, and regional city. This is because in the last 50 years, solar for space and hot water heating has become so widespread. In combination with the appropriate design and construction of new shelters, heating needs have been met with a modest amount of wood, grown for the specific purpose of backing up the solar systems.

In the 20 acres of creek bottom land, people grow such annual monocultures as wheat, corn, rye, barley and oats. Orchards and vegetable gardens are near the houses. Canning of garden products takes place outside, using energy derived from concentrating collectors. Dried foods take precedence over canned foods and root crops are very important.

A single solar hog house on wheels is large enough to accommodate no more than two sows and twenty feeder pigs. A similar solar chicken house, surrounded by a twenty-five-foot square fence, accommodates from twenty-five to fifty chickens. About half are frying chickens, which are eaten during the summer months. Unlike the chickens grown in closed confinement fifty years ago, these animals experience few tumors, and the yolk of their eggs is a brilliant gold.

Pigs and chickens "graze" on fresh pasture during the growing season. Their mobile pens are easily advanced a few feet each day with hand levers operated by school children or grandparents from the village. The mobile pens allow for an economy of fencing materials. This managed migration simulates the migration of large animals in pre-settlement times. Only the breeding stock for pigs, plus the laying hens and two roosters are maintained throughout the winter.

The one large outbuilding is devoted to covering the small amount of machinery. The expensive equipment consists of a small, multiple harvesting combine, a forty-five horsepower tractor and a hay baler. The combine has a seven-foot cutter bar and runs off the tractor's power take-off.

The traction and transportation fuel needs are met with alcohol, derived from crops grown on the farm. The "fuel forty" is the principle energy producer. It grows a six-species polyculture consisting of five grasses and one legume. These species are selected for their high carbohydrate content and relatively low protein yield. This forty-acre field averages about twenty barrels of crude oil equivalent per year. Livestock are cycled onto this acreage for a few weeks each year to enhance the crumb structure of the soil.

Nationwide, roughly 25%, or approximately 100 million acres of cropland, is devoted to growing alcohol fuels. The yield amounts to about 50 million barrels of oil equivalent. An additional 7 million barrels equivalent is gained in the form of methanol from the farm. In 1979, this would have amounted to only a three-day supply of oil, or less than 1% of the annual consumption.

One hundred years ago approximately twenty-five percent of the total acreage was devoted to horses and mules for traction purposes. Now, about eight barrels of crude equivalent, or only ten percent of the total acreage on the farm, is devoted to farm traction. This is because horses and mules would burn energy just standing around being horses and mules, but the tractor can be turned off. However, the tractor cannot become pregnant and build a replacement on solar energy. A pregnant mare at rest is not really resting. Furthermore, parts wear out on the tractor and cannot be replaced by ordinary cell division as with the traction animal. Nevertheless, from the point of view of total energy expenditure, the tractor is used rather than the beast of burden, so long as other livestock are around to enhance the crumb structure of the soil. The other twelve barrels equivalent from the fuel forty, representing about fifteen percent of the total acreage, is sent to the village and city for their liquid fuel needs.

The alcohol fuel "refinery" requires some elaboration. Organic material produced at the farm is delivered to a privately-owned or co-op still in the village. The production of portable liquid fuels is part of a fine-grained approach to our over-all energy needs. It has become economically feasible as farming methods have become less energy-intensive and less capital-intensive. It wasn't economically feasible in the 1980's and produced very little net energy, but the agricultural sector was enthusiastic about producing alcohol fuel from farm crops. Hundreds of on-farm stills were built and closed down in 18 months after federal and state subsidies were withdrawn. Major stills costing $20 million and more were built, and many closed within three years after losing the subsidies. In those years, each automobile would consume in calories what nearly two dozen people would consume in the same period. American farmers learned a valuable and painful lesson about the potential of alcohol fuel production to meet the enormous energy demands of that time. Soil loss accelerated during this period, and farmers gradually learned to curtail their alcohol production programs to a very moderate level.

Another source of energy comes from the "multiple purpose forty." Leaf and stem material are harvested from an herbaceous polyculture after the early summer seed harvest, and converted into methanol equivalent to two to five barrels of crude oil a year. In the fall, some of the net wood production of the woodlot and orchard is also converted. Upon arrival at the still, all organic matter is weighed, moisture is determined, and nutrients are calculated. The farmer may sell some or all of his alcohol into the public sector, but the nutrients left over after distillation are returned to the farm and usually spread on the field or woodlot from which they were taken. Of course human waste is returned to the fields from all over. This is to prevent soil mining and reduce the amount of chemical fertilizer applied.

One concern which is constantly being discussed and fine-tuned in discussions has to do with what tools and equipment should be owned and operated by the farm and which ones made available through the rental place in the village. At this time the rental place provides an Easy Flo fertilizer distributor (for Phosphorus and nitrogen), a chisel which is attached to the tractor to break sod-bound soils, seed bed preparation equipment for the annual crops, plus numerous other pieces of equipment which are not frequently used.

People on this land have a deep distrust of commercially-produced chemicals. It is amazing that this distrust began to develop some forty years ago in the churches. In many seminaries during the 1980's, many students began to discuss the Genesis version of the Creation as possibly contributing to much of the environmental problem. During the 1970's, the dominion question had been much discussed. Since most defenders of the Genesis story had insisted that dominion was not the current word, but that stewardship was implied, church people began to relax. That turned out to be a rather unimportant consideration. During the 1980's another discussion began, much more quietly. The emphasis this time was on the cultural impact of a subtlety in our religious heritage. The culture fostered, however unwittingly, the belief that we are a separate creation. After all, the creation story held that the earth and the living world were created and then there was a pause. Following the pause, in a special effort, came human beings. But our biologists in the last century demonstrated that the same 20 amino acids are in the redwood, the snail, the human and the elm tree, as well as in the lowly microbe. Furthermore, the nucleotides, which make up the code, are mostly the same throughout. Native Americans had talked about brother wolf and sister tree long before these discoveries. Now in our churches it is frequently mentioned that our cells have had no evolutionary experience with such and such a pesticide, or that the concentration of a "natural" chemical much greater than our tissues have ever experienced is to be avoided. A toxic level is defined as a quantity beyond our cells' evolutionary standards.

Because a sustainable agriculture is more important than a highly productive one, upland crops consist of recently developed herbaceous perennial polycultures. The polycultures are ensembles of species developed by the land grant universities through the experiment stations. Perennials were selected because of their soil-holding capability. High-yielding, nutritious seed-producing perennials were first inventoried in numerous experimental gardens. Next, an intense selection program was initiated to increase the yield of the individual species. Later, thousands of species combinations were tried. From then on, plant breeders sought to improve performance of individual species within the polyculture environment. These perennial polycultures have several distinct advantages over the former annual monocultures.

First, soil loss has been reduced to replacement levels. We had expected this, for the reduction of soil loss was a major motivation behind the extensive research. Secondly, spring water has returned to the area. Many springs are now trickling all year around, and the total micro-hydroelectric capacity has increased, along with a rise in the water table. The land with the perennial vegetation has become a huge battery for stored "electricity." A third advantage is that the energy required for maintenance and harvest after the initial planting is just five percent of that required by the former high-yielding annual monocultures. And finally, although the usual pathogens and insects are still around, they do not reach epidemic proportions.

Our particular farm has fields consisting mostly of grasses, a few legumes, and even members of the sunflower family. Some of the fields are harvested in early summer; some, in the fall. The early summer or July harvest in one field includes descendants of Intermediate wheatgrass, Canada wild rye, Sideoats Grama, Tall wheatgrass and Stueve's Lespedeza. The fall harvest consists of four grasses, a legume and a member of the sunflower family. The grasses include descendants of switchgrass, sand lovegrass, Indian Grass, and weeping lovegrass. The legume, Wild Senna, and a perennial soybean provide nitrogen and some seed, and a high-yielding descendant of the Gray-headed Coneflower produces seeds with two important oils.

Some of the early objections to the difficulty of harvest and separation of seeds from the polycultures were quickly overcome when agricultural engineers began to invent new machinery. In fact—and this is ironic—the return to polycultures only became possible in our age of mechanization. Some have since made the argument that monoculture arose because of the need to harvest small seeds efficiently when all we had was hand labor. The age of mechanization, then, has allowed us to develop an agriculture that closely mimics the vegetative structure in pre-agricultural times. Much of the machinery has allowed us to return to the psyche of hunters and gatherers again, but of course in a modern context.

The fossil fuels used during the transition era, 1985–2025, as we moved from mining and destruction of land as a way of life to the solar age, afforded us opportunities not only in plant breeding, but in animal improvement as well. This period gave us the chance to develop crops that were less dependent on humans. The same was true with the livestock. For example, American Bison were crossed with domestic cattle and the thicker hides made the critters more resistant to severe winters.[2] In a way, we are now using solar energy (stored in grass) to maintain barns—the hides of animals. Protective shelter made of lumber for large animals is not necessary.

Grandparents amuse children with stories about square tomatoes and featherless chickens. The featherless chicken was developed in the 1970's by reductionistic technologists who thought they would help corporate chicken growers and processors to cut costs in cleaning chickens. The consequence was a funny-looking chicken that required such a warm environment that the energy costs were in excess of the cleaning costs. The moral of the story is that big money is a sure license for big foolishness.

Livestock are moved from one polyculture to another in a rhythm that does not jeopardize flowering and seed set. Most grazing does occur in areas that produce seed for human consumption, but certain polycultures are grown for the livestock exclusively. A few weeks before slaughter, buffalo/beef graze on mixed perennials that are setting seed. This is a weak simulation of the feedlot of former times. No hay is hauled to the barn for winter feeding, for there is no barn. Some of the hay is windrowed with a side-delivery rake and left, but most raked hay is rolled into 1000–1200 pound bales and left in the field. This system reduces the need to spend time and energy moving hay and manure, and the nutrients are left where they are most useful.

The movement of livestock on the farm turns out to be critical. In natural ecosystems there were no fences. Even though we are forced to use fences for all our livestock, our management program recognizes that animal wastes on the farm contribute to the crumb structure of the soil, which allows the soil to release nutrients slowly while holding moisture.

Many of the problems caused by farming techniques of the 1960's and 1970's have been solved in this new era. Seed bed preparation occurs mostly where the few acres of annuals are grown, and since tillage has been dramatically reduced, soil loss is almost non-existent. Silting of streams is minimal, and more species and larger populations of fish thrive in the waterways. Energy-expensive terracing is no longer as necessary, and where check dams and small farm ponds exist, they serve the farmer mostly as pools for growing catfish. Irrigation is reduced, because hundreds of springs have been reborn. Fertilizer application is minimal because the diversity of crops has maintained a better nutrient balance with less nutrient run-off. The record-breaking fish kills because of fertilizer and feedlot run-off during the last century are now only part of the legends about our unenlightened grandparents. Weeding is essentially a thing of the past, except in gardens and where annual monocultures are grown. Pesticide application is almost non-existent because of both polyculture and a broader genetic base in our crops. A broader genetic base in livestock and the demise of high-density feedlots have made the use of antibiotics for livestock seldom necessary. The life of farm machinery has increased by a factor of sixteen in the last fifty years. All of these changes have resulted in a drastic cut in energy consumption for farm production.

The major changes began to surface during the 1980's, when a few young agricultural professionals, having adopted a sustainable agriculture as their goal, looked for the sustainable alternatives rather than placing their bets on corporately-controlled agriculture. In many respects, they were the true heroes of the era. Some took the theory of the quantitative gene developed during the 1960's and, along with the known virtues of hybrid vigor, made repeated breakthroughs in new crop development.

There was a unifying theme from Massachusetts to Kansas to California. People recognized that in the long run and often in the short run, land determines. Citizens sought to meet the expectations of the land and to look at the natural ecosystems of different regions as the standards against which to judge their agricultural practices.

Suddenly, as is so often the case with profound statements, there was a new meaning to the words Thoreau uttered from the Concord Lyceum in the mid 1800's: "In wildness is the preservation of the world."

The policy-makers began to take seriously the prediction of Charles Lindberg: "The human future depends on our ability to combine the knowledge of science with the wisdom of wildness."

When this concept was applied to our farms, they became waterproof, diversified family hearths. Our fields are no longer vulnerable, oil-hungry monocultures, although they are not wilderness either.

But without wilderness, we would not have developed a sustainable agriculture and culture. The practice of stewardship is now both easy and effective; for plowshares have been beaten into appropriate tools and war against our Earth Mother is practiced no more.

Reference

1. Henry A. Wallace, 1934. *New Frontiers,* Reynal and Hitchcock, New York. pp. 274–275.

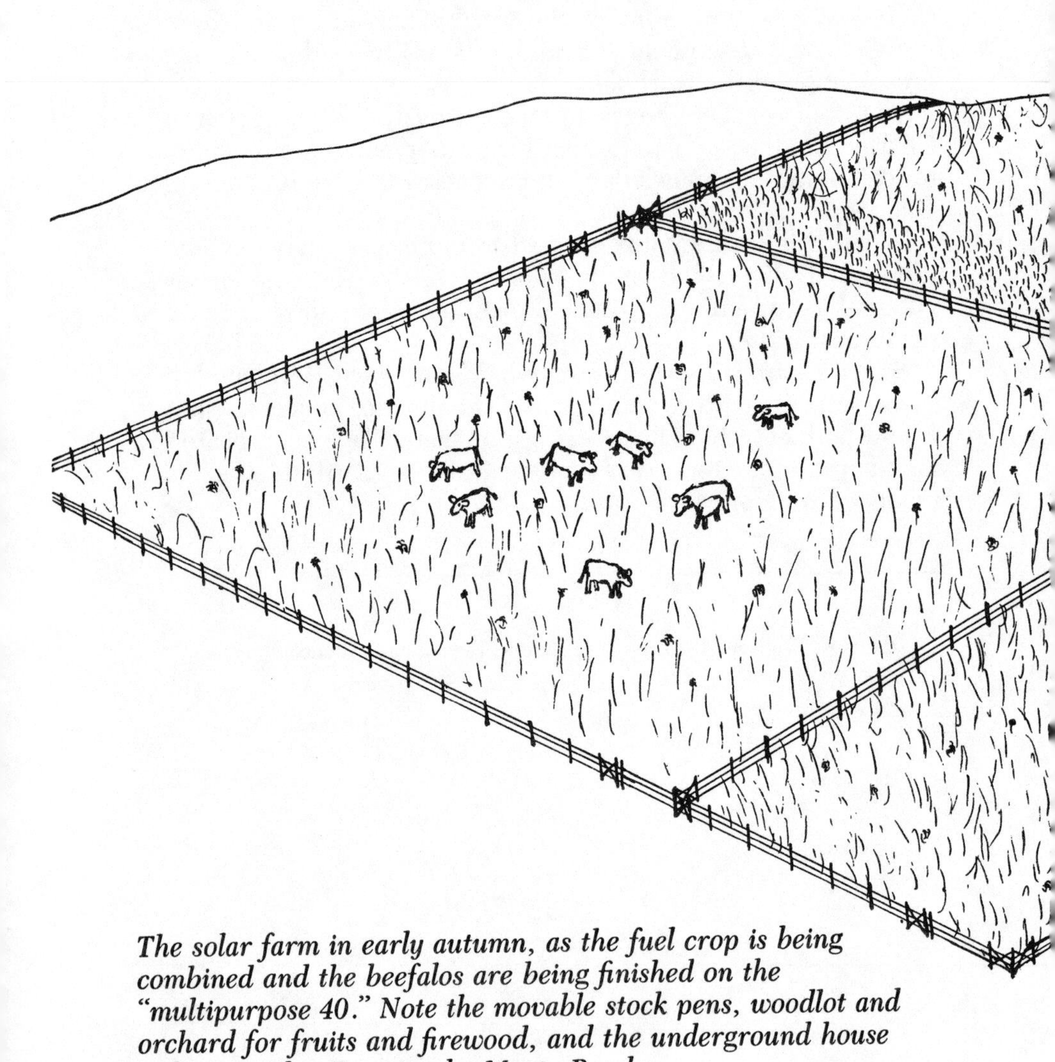

The solar farm in early autumn, as the fuel crop is being combined and the beefalos are being finished on the "multipurpose 40." Note the movable stock pens, woodlot and orchard for fruits and firewood, and the underground house at lower right. Drawing by Marty Bender.

MB

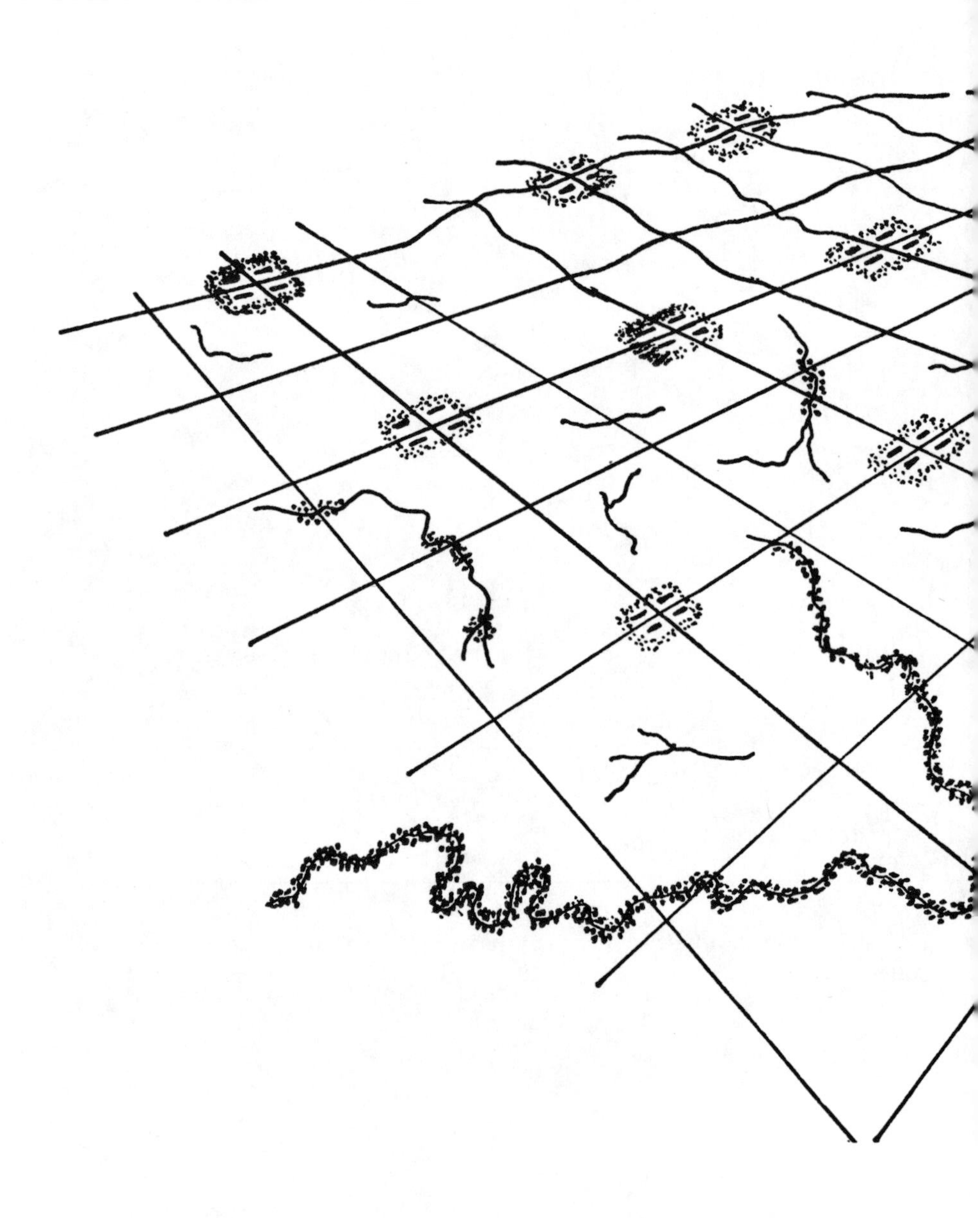

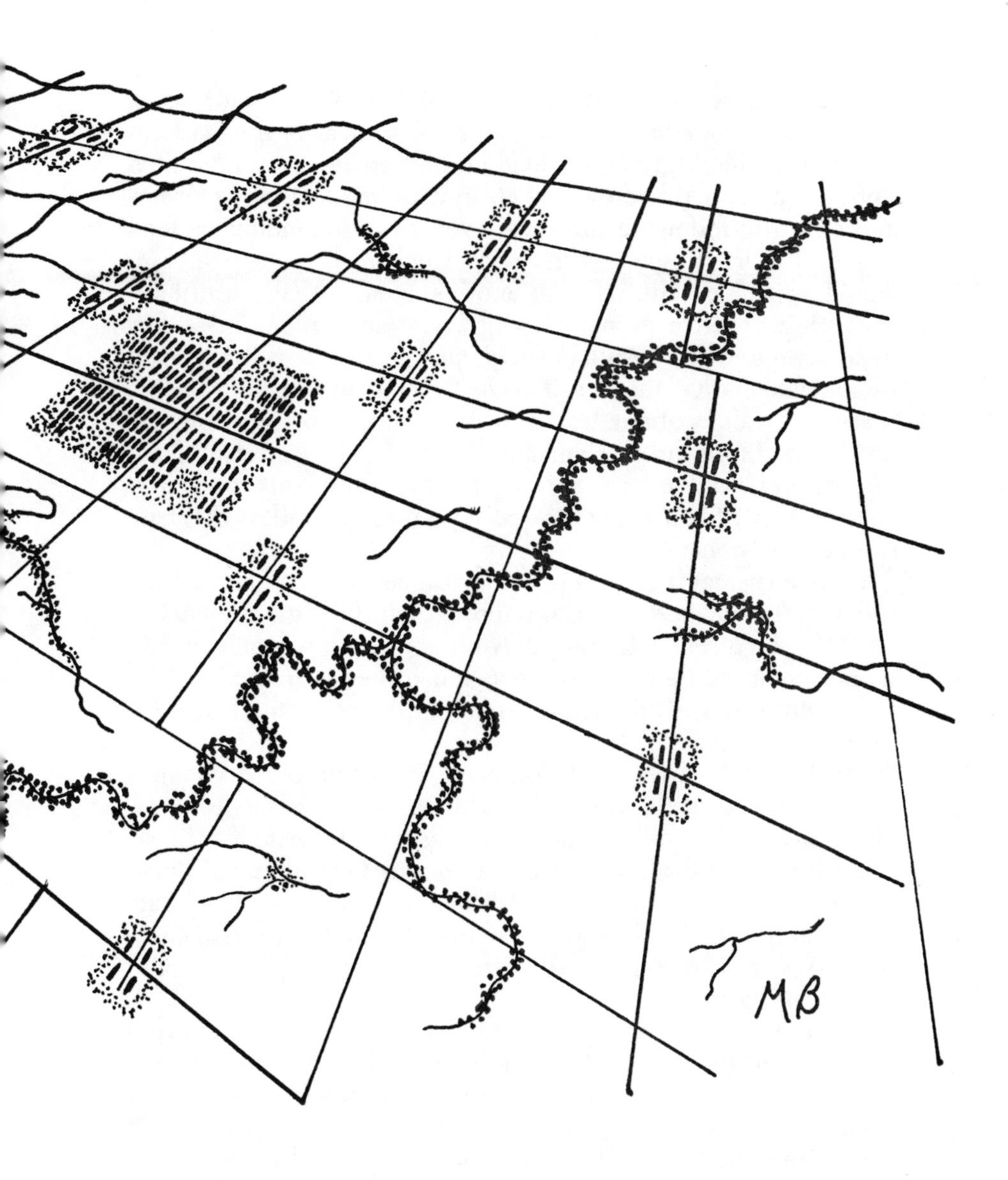
MB

Acknowledgements

This book began as much in the kitchen of my youth as in the fields. In our farm kitchen scarcely a meal began without a prayer of thanksgiving for food and other blessings or ended until the plate had been wiped clean with bread. During dishes and clean up following the meal even so little amount of food as half of a fried egg would be returned to the refrigerator; it would appear on the table at another meal, usually in a hash consisting of left-over potatoes, gravy, some onion and a few meat scraps. This wasn't poverty, just frugality well-managed by an imaginative mother. I have become increasingly aware that more values about land and its relationship to people are taught in dining areas than anywhere else in America. I am forever indebted to my parents for providing this treasured environment, for it concentrated my mind on land as the true source of sustenance and health.

I owe a special thanks to my wife, Dana, who read and edited this manuscript and made numerous helpful suggestions. Because we co-direct activities at The Land, both officially and in reality, many of these ideas are as much hers as mine. She has also been a source of encouragement and a sounding board for many crazy ideas.

My friend Charles A. Washburn, Professor of Mechanical Engineering at California State University, Sacramento, provided me the first substantial encouragement to write this book. Later he read a draft and made numerous helpful suggestions. During the summer of 1979, at The Land, he assembled data and talked with local farmers on the alcohol-from-grain question. Most of Chapter 6 is the result of his early questions and our collaboration.

Students and research associates at The Land could not avoid contributing to this effort, for ordinary politeness often forced them to listen. Research Associate Marty Bender and I have spent countless hours discussing and thinking about the "soft agricultural path." He has called numerous references to my attention and has critiqued the major ideas. I owe him special thanks. Other special thanks are due Mari Peterson who collaborated with Charles Washburn and me on an alcohol fuels paper. Mari has always given an intelligent response to considerations discussed in this book. I have appreciated Kelly Kindscher's presence at The Land during the 1979–80 school year. His investigations on ecological agriculture should make a fine contribution to those seeking an alternative agriculture for the near future. He has raised numerous important questions about the central idea of this book.

I am grateful to all Land Institute students who have made a contribution to the thinking in this book, but those interested in alternatives in agriculture deserve a special thanks: Michael Chapman from the University of California at Santa Cruz; Bill Craig from Kansas State; Mark Lieblich from Rutgers; and Karl Zimmerer from Antioch in Ohio can all be found somewhere in the ideas presented here.

Professor Major Goodman at North Carolina State University read much of Chapter 10 and made comments that helped, even though I am sure he remains skeptical about the potential of perennial grain crops.

Thanks are also owed to Amory Lovins, Dave Brower and Bruce Colman of Friends of the Earth. Lovins encouraged me to write the book. Brower insisted that I write it and Colman was my editor. Without such an effort, I am confident the manuscript would never have made it to the printer.

Finally, I thank Pam Ellinghausen, both for her patience and for typing the manuscript, much of it twice. I also thank my daughter, Laura, for typing and reference checking. Both Pam and Laura caught errors and noticed some awkward sentence structure.

Scores of others have contributed to this book and I would mention them were I not afraid of missing some important contributors. There are many good people in the environmental movement who deserve special recognition. I would like for them to share in the errors which may be found in this book—but of course that responsibility must rest with me.

—Wes Jackson

About the Author

Wes Jackson, director of The Land Institute, was born in 1936 on a farm in the Kansas River Valley near Topeka. He holds a B.A. in Biology from Kansas Wesleyan, an M.A. in Botany from the University of Kansas, and a Ph.D. in Genetics from North Carolina State University. After teaching high school and at Kansas Wesleyan and California State University at Sacramento, in 1976 he founded The Land Institute, of which his wife, Dana, is a co-director.

His previous book is *Man and The Environment* (W.C. Brown, 1971), a collection of readings which has been widely adopted by college environmental survey courses.

About The Land Institute

The Land Institute lies on 28 acres on the Smoky Hill River, five miles south of Salina, Kansas. It is a non-profit educational research organization, devoted to the study of sustainable alternatives in agriculture, energy, waste-management, and shelter. Its small school building and study area is powered by a pair of wind electric systems and is heated by solar collectors, a solar greenhouse and wood stoves.

The Institute publishes *The Land Report* three times a year, under the editorship of Dana Jackson, and takes in 8 to 10 graduate and undergraduate students each semester. The *Report* writes up student activities, research projects, and special events.

The Land Institute is supported by tuitions, grants and private gifts. For further information, write:

The Land Institute
Rte 3
Salina, Kansas 67401

ABOUT FRIENDS OF THE EARTH

We invite you to join Friends of the Earth, a dynamic environmental group. We work on the international scale to promote the most worthwhile option there is for the future—a living planet.

With national groups in 14 countries, active lobbies in Washington, DC and other capitals, and an aggressive publishing program of books and periodicals, we use every legal means at our disposal to promote changes that will allow Man to live in balance with this, our only planet.

All members receive *Not Man Apart* twice a month and are entitled to discounts on FOE books. Members in categories of $35/year and over receive FOE gift books as premiums.

Dues in FOE are not tax-deductible, in order that we may lobby vigorously.

Friends of the Earth Foundation solicits deductible contributions of money or property.

EARTHWORKS

Ten Years on the Environmental Front

by the staff of Friends of the Earth

edited by Mary Lou Vandeventer

Wit and insight have been the hallmarks of Friends of the Earth's newspaper. Since 1970, *Not Man Apart* has brought out the vital stories, calls for action, and thoughtful essays and reviews.

Earthworks presents the best of *Not Man Apart,* including humor, poetry, think-pieces, and the full range of the last decade's environmental news on energy, nuclear power, pollution, economics, alternative technologies, wilderness, education, and wildlife.

Not just a look back at the decade that began with Earth Day, nor a retrospective of the best in environmental writing, *Earthworks* is an anthology of the literature and art that have given environmentalism the political force to improve people's lives.

What readers say about Not Man Apart:

"A must for any library. The tone is moderate, yet firm; the photographs and illustrations good, and the overall quality rather exceptional." —*Library Journal*

". . .covers the eco-front with passion and objectivity—the happy combination."

—Garrett Hardin

Contributors:
Edward Abbey • Barry Lopez • David Comey • Joan McIntyre • Josephine Johnson • Wendell Berry • Edward Hoagland • David Gancher • Jim Harding

ISBN: 0-913890-39-1
trade paperback, illustrated, 224 pages
$8.95

THE NEW ENVIRONMENTAL HANDBOOK

edited by Garrett DeBell

The New Environmental Handbook follows the spirit of the book that kicked off the original Earth Day movement. Equipped with the *New Handbook*, the reader will be able to help solve the energy crisis, clean up pollution, preserve wilderness, and take part in the big public campaigns of the 1980s.

Garrett DeBell's team of writers and specialists looks at the gains we have made in the last ten years and at emerging challenges and opportunities: Amory Lovins's soft energy paths, for example; the evolving fields of self-help and preventive medicine; transportation options; and widgets that improve car mileage, save water in the home, or reduce utility bills.

The issues are put into brief and concise perspective, with a focus on what can be, and why, and on what you can do to bring it about.

Contributors:
David Brower • Paul Ehrlich • Lynn White, Jr. • Garrett Hardin • Kenneth Boulding • Amory Lovins • Mark Terry • Ross Pumphrey • George Alderson • John Gofman • Denis Hayes • Senator Durkin • Senator Proxmire

ISBN: 0-913890-36-7
trade paperback, 168 pages
$5.95

solar architecture

PRESENT VALUE

Constructing a Sustainable Future

by Gigi Coe

foreword by Denis Hayes

introduction by Wilson Clark

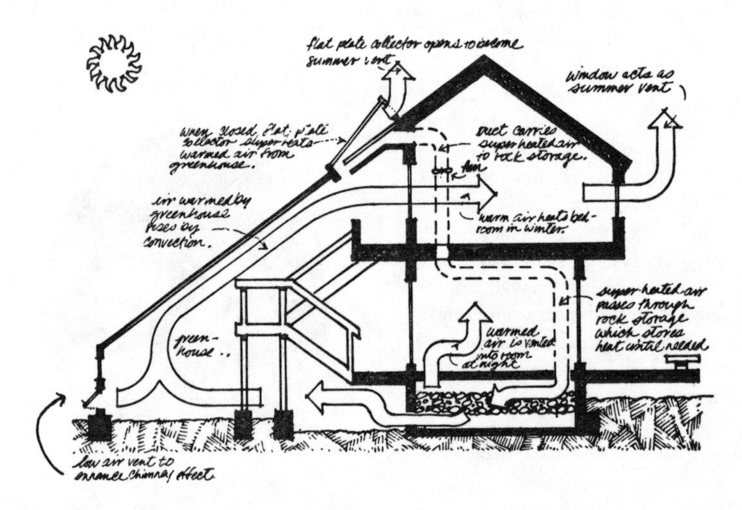

Proof positive that solar energy works! In the 20 buildings featured in *Present Value,* people use solar energy every day, economically, to run homes, offices, factories and farms.

Present Value shows how these solar energy systems work, how much they cost to install, and how much they save on fuel and utility bills. Architectural drawings, photographs, and text illustrate the sun heating—and cooling—these work and living spaces, with both active and passive systems.

Present Value proves that the sun is an economical energy choice in conditions ranging from snowy mountain winters to foggy coasts, to dry desert summers.

"We must learn to live what we believe in. . . We can all learn from the experiences described in this fascinating book."

—Solar Energy Research Institute

ISBN: 0-913890-35-9
trade paperback, illustrated, 96 pages
$5.95

where to find it

THE ENERGY AND ENVIRONMENT CHECKLIST

by Betty Warren

Do you want to build a solar-powered home? Fight a nuclear power station? Do most anything in between?

The *Checklist* has some fourteen hundred annotated entries that will tell you where to find the information you need.

Films, books, slideshows, multi-media presentations, magazines, and organizations concerned with energy and environment are evaluated under two dozen category headings.

A typical entry tells the author, title, publisher or distributor or sponsor, length, technical level, and price of the resource—and where you can buy it by mail order.

This is the third revision of FOE's popular *Energy and Environment Bibliography*, a unique and popular resource.

ISBN: 0-913890-37-5
paperback, 100 pages
$5.95

- ☑ Transportation
- ☑ Citizen Action
- ☑ Utilities
- ☑ Wind Energy
- ☑ Nuclear Energy
- ☑ Graphics
- ☑ Catalogs
- ☑ Sources of Information
- ☑ General References

To: Friends of the Earth
124 Spear Street
San Francisco, CA 94105

Please Join Us.

☐ Please enroll me for one year in the category checked, entitling me to *Not Man Apart* and discounts on selected FOE books.
(*Contributions to FOE are not tax-deductible.*)

☐ Regular = $25	☐ Spouse = add $5
☐ Supporting = $35*	☐ Life = $1000***
☐ Contributing = $60**	☐ Patron = $5000***
☐ Sponsor = $100**	☐ Retired = $12
☐ Sustaining = $250**	☐ Student/Low Income = $12

*Will receive free a paperback volume from our *Celebrating the Earth* Series.
**Will receive free a volume from our *Earth's Wild Places* Series.
***Will receive free a copy of *Headlands* (our award-winning, gallery-format book).
☐ Check here if you do not wish to receive your bonus book.

☐ Please accept my *deductible* contribution of $ ________ to Friends of the Earth Foundation (*checks must be made to FOE Foundation*).

Please send me the following FOE Books:

	Cost
________________________________	____
________________________________	____
________________________________	____
Subtotal	____
6% tax on Calif. delivery	____
Plus 5% for shipping/handling	____
TOTAL	____

☐ Send full FOE Books catalogue.
☐ VISA ☐ Mastercharge
Number ________________ Expiration date ________
Signature ________________________________
Name ________________________________
Address ________________________________
City ________________ State ________ Zip ________